PRENTICE-HALL BIOLOGICAL SCIENCE SERIES

William D. McElroy and Carl P. Swanson, *Editors*

Concepts of Modern Biology Series

CONCEPTS OF MODERN BIOLOGY SERIES

William D. McElroy and Carl P. Swanson, Editors

CLARK P. READ

Professor of Biology
Rice University

Animal

Parasitism

PRENTICE-HALL, INC., Englewood Cliffs, New Jersey

10 9 8 7 6 5 4 3 2 1

ISBN 0-13-037663-9   P
      0-13-037671-X   C

Library of Congress Catalog Card Number: 72-39331

Printed in the United States of America

PRENTICE-HALL INTERNATIONAL, INC., *London*
PRENTICE-HALL OF AUSTRALIA, PTY. LTD., *Sydney*
PRENTICE-HALL OF CANADA, LTD., *Toronto*
PRENTICE-HALL OF INDIA PRIVATE LIMITED, *New Delhi*
PRENTICE-HALL OF JAPAN, INC., *Tokyo*

chapter 1   A Perspective   1

chapter 2   Protozoan Parasites   9

chapter 3   Trematode Parasites   54

Contents

chapter 4   Tapeworm Parasites   80

chapter 5   Acanthocephala   105

The main body of biological literature consists of the research paper, the review article or book, the textbook, and the reference book, all of which are too often limited in scope by circumstances other than those dictated by the subject matter or the author. Unlike their usual predecessors, the books in this series, CONCEPTS OF MODERN BIOLOGY, are exceptional in their obvious freedom from such artificial limitations as are often imposed by course demands and subject restrictions.

Today the gulf of ignorance is widening, not because of a diminished capacity for learning, but because of the quantity of information being unearthed, most of which comes in small, analytical bits, undigested and unrelated. The role of the synthesizer, therefore, increases in importance, for it is he who must take the giant steps, and carry us along with him; he must go beyond his individual observations and conclusions, to assess his work and that of others in a broader context and with fresh insight. Hopefully, the CONCEPTS OF MODERN BIOLOGY SERIES provides the opportunity for decreasing the gulf of ignorance by increasing the quantity of information and quality of presentation. As editors of the Prentice-Hall Biological Science Series, we are convinced that such volumes occupy an important place in the education of the practicing and prospective teacher and investigator.

WILLIAM D. MCELROY

CARL P. SWANSON

Series Foreword

To Lee, Jo, Steve, Vic, Jeff, and Cathleen

As a biological phenomenon, parasitism has in the past been a very traditional subject included in the education of undergraduates committed to biology. However, in most of the books concerned with the subject, the treatment of animal parasitism has been strongly oriented to medicine or veterinary medicine. As such, the study of parasitism has suffered considerable neglect as an old-fashioned subject. Also, the subject has seemed less attractive to students of biology because they felt that scholars in the field did not include treatment of molecular events and tended to neglect the mainstream of biology.

This brief book has therefore been written with the view that parasitism should be treated as a legitimate area of biology and that the development of this particular field of study is a relevant subject for the consideration of biologists. The study of animal parasitism has had a distinguished history, and the author believes that it has a distinguished future. It is hoped that this book will present a short integrated view of the area of animal parasitology and will also attract students with interests in such diverse fields as evolution, genetics, biochemistry, and developmental biology to the use of host-parasite systems for the investigation of fundamental problems.

No real attempt has been made to deal in this book with symptoms, diagnosis, or treatment of parasitic disease; it is not a reference book for clinicians. However, it may be valuable for students who want to be clinicians in the future because it attempts to treat parasitism in a broad biological context. This is important since the ultimate relevance of this subject involves human welfare.

Preface

CLARK P. READ

Houston, Texas

Said the Duchess, "Everything has a moral if you can find it."

—from ALICE'S ADVENTURES IN WONDERLAND

by Lewis Carroll

The study of animal parasites has had a long and distinguished place in the history of biology. The larger animal parasites, such as *Ascaris* or the tapeworms, were probably the first infectious agents seen by man. The invention of the microscope was, of course, necessary for the recognition of the great world of organisms living below the range of the unaided eye. However, descriptions in Egyptian papyri, in the Old Testament, and in ancient Chinese literature leave little doubt that parasitic animals such as ascarids, hookworms, Guinea worms, and perhaps blood flukes were known to ancient scientists. Certainly some of the diseases associated with these parasites were recognized and described, although the connection between disease and parasite may not always have been apparent to these early observers.

Parasitic animals were favorite subjects for study in resolving the controversies concerning the spontaneous generation of living things. Pasteur's classical study on an infectious disease of silkworms involved a protozoan parasite. The modern concept that specific chemotherapeutic agents against disease organisms might be feasible was derived from studies on parasitic flagellates. The first observations on the phenomena of drug resistance were made on parasitic animals. The eggs of animal parasites long served as a classical material for the cytological events of early embryogenesis. Much early descriptive ecology had its origins in the attempts of biologists to unravel the details of the life cycles of parasitic animals. Thus for a long period of time animal parasitology was firmly entrenched in the matrix of biological science.

The study of animal parasites as a legitimate area of biology has fallen into considerable disuse in the past few years. This is attributable to several causes:

A
Perspective

chapter 1

1. The tendency of writers to treat animal parasites in a highly traditional framework, leaving the discussion of functions, that is, the physiology and biochemistry, of the organisms, to specialists who write reviews to be read by other specialists.

2. A breaking-up of parasitology as a professional field, with more and more young workers identifying their researches on parasites with comparative biochemistry and physiology, developmental biology, cell biology, and so forth. (This, in turn, has led to further scattering of scientific data in the research literature.)

3. The failure of parasitologists to bring their field into the mainstream of biology. This is not to say that the study of biology should be a popularity contest. However, the advances in biochemistry, cell biology, and population biology have not been adequately integrated into the reframing of questions concerning animal parasites. Further, the *use* of parasitic animals in answering more general questions of biology has rarely been exploited. The use of bacteriophages for the examination of general questions in molecular biology should have served as a clue to the possibilities of host-parasite systems as tools in answering questions of general significance in biology, but this clue has been disregarded. Parasitism should be examined as an adaptive control system at several levels of organization.

4. The general failure of communication between fields of biology dealing with the phenomenon of parasitism. For example, the transmission of information, or even of concepts, between research workers in animal parasitology and those in plant pathology has been minimal, although each of these groups has much to offer to the other.

The decline of the romanticism in parasitology has been clear. The brave tales of Ehrlich, Manson, Bruce, Theobald Smith, and other pioneers of medical and veterinary parasitology were exciting in the historical sense, but they did not seem personally relevant to many bright young minds of the fifties and sixties. This is as it should be. But, then, what—or who— is taking the place of these men who sought to unravel the mysteries of the diseases resulting from animal parasitism? The fact remains that many of these diseases have not yet been conquered, although some of them have been pushed out of the temperate zones. Apparently a changing geopolitics has effectively modified European and American interest in the tropics, and many governments have lost interest in parasitology.

All of the above may seem quite disheartening, but this need not be the case. Rather, it can serve to emphasize the fact that the study of parasitism is an underdeveloped area of biology that is ready for the attention of the talented and the curious. It has been said that phenomena that appear to be widespread and obvious are likely to be important. In that case, parasitism is indeed important. Numerically, more organisms live in a close physiologically dependent relationship with other species—that is, *in* other organisms—than live what has been termed a "free-living" existence. A search of nature reveals that there is no such thing as an uninfected host. Even the so-called germ-free animals reared in the laboratory are usually infected with one or more viruses. A life style that is so widespread must have significance beyond that of a natural history or a cabinet of curios. Some writers might refer to the study of animal parasitism as a specialized subject. On the contrary it is a very broad area of ecology and calls for the conceptual application of physiology, biochemistry, cytology, genetics, general ecology, systematics, and systems analysis. It is basically multidisciplinary in character.

I believe that we are now in the initial portion of a new phase of parasitology. There are stirrings that indicate that a new integration of animal parasitology with other areas treating of parasitism, plus the application of new methods and concepts from molecular and cell biology and from population ecology, may revamp the field. Although there is no doubt of the pragmatic value of parasitology in human and animal health, there now appears to be reason to expect that the study of parasites and parasitism may yield new fruit of broad general significance. We have already mentioned the importance of studies on bacteriophages to general biology. Other parasites offer different kinds of opportunities for the solution of problems. As we shall see, there are certain animal parasites that have highly abbreviated metabolisms and simplified regulatory systems, which offer advantages as experimental material; and there are those that show specialized and readily manipulated control systems for morphogenesis.

Therefore, although there may be some readers who feel that this book is too late or premature, it is hoped that an abbreviated treatment will present some interesting knowledge and views of animal parasitology and lead to interesting biological problems for which animal parasites are highly suitable experimental material.

WHAT IS PARASITISM?

If the average educated person is asked to define parasitism, the response is likely to be couched in terms of disease and to be accompanied by expressions of distaste. This is attributable to the fact that great emphasis has been placed on diseases as manifestations of parasitisms involving man and his domestic animals. However, it should be carefully pointed out that although disease is often indeed a manifestation of parasitism, a definition of parasitism must be phrased in ecological terms. Parasitism is a way of living in which an organism, the parasite, uses an organism of a different species, the host, both as a habitat and as food. It is a special way of living within a broader ecological category known as *symbiosis,* a term used by de Bary in 1879 to describe organisms of different species living together in intimate association. The essence of the differentiation of parasitism as a category of symbiosis lies in the sometimes intuitive recognition that the parasite grows and reproduces at the expense of the host. In terms of thermodynamics a decreased entropy of the parasite entity is typically accompanied by an increase in entropy of the host entity. Disease itself may be defined as an increase in the entropy of the host steady state.

It is perhaps important to distinguish parasitism from predation. The British ecologist Elton remarked that predators live on capital and parasites live on income. In a traditional examination of an ecosystem it is recognized that there is a decrease in numbers of individuals and of biomass as we move from primary producers (plants) to herbivores to carnivores. The well-known pyramid of numbers is an expression of this concept (Fig. 1–1). However, parasites do not fit into the pyramid of numbers along with the carnivores. There are more individual parasites than there are hosts. This is attributed simply to the fact that the host ("prey") is the habitat. The parasites, on the other hand, would lie at the apex of a pyramid of biomass or a pyramid of numbers (Fig. 1–1). Kormondy (1969) may be consulted for a general discussion of such relationships.

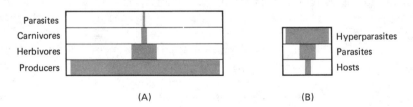

Fig. 1–1. The ecological pyramids of biomass (A) and of numbers of organisms (B).

It is important to recognize that the term *parasitism* includes both host and parasite, since the parasite removed from its habitat is no longer parasitic. It has been hard for biologists to think in these terms. A major difficulty lies in the cybernetic aspects of parasitism. The habitat (host) of the parasite is responsive, not only to the external environment but also to internal environmental changes that may be produced by the parasite. It should be obvious that a deep understanding of the operation of a given parasitism requires a special understanding of the physiology of both host and parasite. It is not sufficient to understand the function of the host and parasite as separate entities, but rather it is necessary to examine the nature of responsiveness in the components when host and parasite exist as a biological entity. In systems terms the "host parasite" is the expression of host and parasite interaction, and as a system it differs from its parts.

From a purely practical view, how can parasitism be recognized in nature and distinguished from other forms of symbiosis? It must be answered in truth that in many cases parasitism cannot be recognized with certainty. When disease is an obvious result of a host-parasite combination, this may be construed as evidence of parasitism. However, an examination of any group of animals will reveal that apparently healthy animals frequently serve as hosts to animal symbiotes. It is often virtually impossible to determine whether the symbiote and host combination is a parasitism or some other form of symbiosis. One rule of thumb often used by parasitologists is to assume that a symbiote is a parasite if it belongs to a taxonomic group in which related forms are known to be parasites. A difficulty in the application of this criterion is that in many instances the related forms are *not known* to be parasites.

From what has been said it will be seen that the study of parasitism is a special type of ecology. It is special in terms of the host-parasite physiological relationship. On the other hand, parasitism may be examined in the framework of ecology in a quite different sense. The determinants of the rate or frequency of transmission of parasites from one host to another, including the complications of the life histories of the parasite and host when they are not together, constitute subjects for a more traditional ecology. Those aspects of weather, for example, that determine whether a parasite lives or dies upon leaving the host are features that may be treated in quite a different way than the physiology of the host-parasite relationship. Insofar as understanding of natural phenomena is concerned, these aspects are interlocking components of operating biological systems, and their rigorous separation is a confession of human limitations. Biologists may limit their horizons in one mode, but other organisms live under other classes of constraints.

Parasitism is a broad term and includes a great many relationships that cannot be considered in the present context. Although we shall be concerned with parasitic animals, the majority of parasites belong to the groups of organisms that are microscopic or ultramicroscopic in dimensions. These, of course, are the viruses, bacteria, and fungi. The fields of study that have grown up around all these parasites and parasitisms are highly diverse in origin and in orientation. Plant pathology, for example, grew from agriculture, whereas bacteriology and the study of bacterial diseases of animals developed as a separate discipline focused in medicine and veterinary medicine. Animal parasitology received its greatest stimulus from European invasions of the tropics and the development of tropical medicine.

Therefore, men being men first, and scientists secondarily, it is not surprising to find that the taxonomy of science itself and the segregation of disciplines or craft guilds have interfered with the transmission of information between those knowledge areas that have in common a concern with parasitism. Although we cannot treat this subject of parasitism in a very broad context in this little book, the point is made in order to emphasize the need for perspective in dealing with phenomena of this breadth. This last statement is based on the concept that we are indeed talking of related phenomena when we speak of parasitism. We shall attempt to examine this idea as it may apply to parasitic animals, but some of the conclusions may be quite applicable to other groups of parasites.

PHYLOGENETIC CONSIDERATIONS

We have already remarked that there are numerically more organisms living inside someone else than there are organisms living the so-called free life. When we examine the phyla of animals, it is striking to note that although there are parasitic representatives of most of the phyla, parasitism occurs more often in some groups than in others. This mode of life is relatively common among the Protozoa, the flatworms, and the nematodes, and it is fairly common among the arthropods. Further, it is also apparent that the adoption of a parasitic mode of life has occurred independently within these groups. In the nematodes and arthropods, for example, it has clearly occurred on a number of independent occasions. The same thing may be said of the protozoans, although the phylum Protozoa is, in some ways, not really comparable to the phyla of multicellular organisms.

Since parasitism has arisen on many separate occasions, it is legitimate to pose the questions of whether such organisms have attributes in common and whether any general principles seem to pertain in the phenomena of parasitism. It is obvious that there are certain characteristics that will not be held in common by animal parasites. Since the evolutionary origins are diverse, the constraints imposed by phylum morphology means that the organisms will not look alike. Thus we might expect to examine physiological and chemical aspects of animal parasitism in a search for principles that might govern or be associated with the adoption of parasitism. The constraints of phylum origin will also be apparent in the pattern of the life cycle available to an organism. For example, nematodes typically molt during the life cycle. It is therefore not surprising to find that the parasitic nematodes molt. This is an evolutionary physiological constraint, and we must recognize that there are evidently levels of organization and function that cannot be tampered with if the organism arose from an ancestor that already possessed constraints imposed by evolution.

"PROPERTIES" OF THE PARASITE

Looking at the question in the broadest possible way we can, however, recognize certain features held in common by parasites. In parasitism the parasite lives in its food supply and feeds upon it. It must have adaptations for recognizing and reacting to the presence of a host. It must have mechanisms for getting from one host to another. We can systematize this a bit by a taxonomy that would include the following:

1. *Infectiousness:* This would encompass mechanisms for tolerance of the peculiarities of instant introduction to a new environment. The ability of a parasite to resist the action of free hydrochloric acid in the stomach of a vertebrate or to withstand the elevated temperature in a bird or mammal illustrates this point. The shock of this kind of change is obviously instantaneous. Further, the parasite must react to this new environment in order to become established.

2. *Establishment:* This entails mechanisms for quickly shifting to a state of physiological effectiveness, allowing growth in a new environment (frequently involving such factors as anacrobiasis and nearly always involving resistance reactions of the host) and the interlocking of parasite physiology with the homeostatic mechanisms* of the host.

3. *Transmission:* Reproductive function must be linked to adaptations for reaching a new host individual. Since the environment furnished by a host is an ephemeral entity, reproductive rate must be high, and adaptations for transmission must frequently be elaborate. (Although superficially such adaptations seem simple in certain cases, close examination always reveals otherwise.) This may include activity in some other host or in an environment outside a host.

Rather than discussing in what would have to be very general terms the ways in which parasites exhibit the special characteristics just outlined, it seems better to choose first a few exemplary organisms and dis-

---

* An old term for the feedback mechanisms involved in the maintenance of a steady state.

cuss them in somewhat more detail so as to show how these characteristics may be manifested. It is quite impossible to consider general problems of parasitism, or anything else of importance, without a framework of facts with which to work. Some degree of competence is a requirement. There was a tendency for some university students of the 1960s to say, "Don't bug me with details. Just fill me in on the big picture." Unfortunately, the "big picture" is quite blurred without some details. The meaningful trick is to distinguish the significant from the trivial. I have tried to select those organisms about which the significant features are known, but a few trivia may have been included to round out the picture, and some important details may have been inadvertently omitted.

At the end of the book we will return to the question of whether any generalizations can be made concerning parasitism.

REFERENCES

Brand, T. von. 1966. *Biochemistry of Parasites.* New York: Academic Press, Inc.

Fallis, A. M. (ed.). 1971. *Ecology and Physiology of Parasites.* Toronto and Buffalo: University of Toronto Press.

Foster, W. D. 1965. *A History of Parasitology.* London: Ballière, Tindall & Cox.

Kormondy, E. J. 1969. *Concepts of Ecology.* Englewood Cliffs, N.J.: Prentice-Hall, Inc.

Read, C. P. 1970. *Parasitism and Symbiology.* New York: The Ronald Press Company.

Rogers, W. A. 1962. *The Nature of Parasitism.* New York: Academic Press, Inc.

Trager, W. 1970. *Symbiosis.* New York: Van Nostrand Reinhold Company.

Unicellular eucaryotic organisms seem to live as parasites in all species of multicellular organisms, and some even parasitize other unicellular organisms. As might be anticipated, some of them live intracellularly during a major portion of the life cycle, and others live in extracellular locations such as body fluids, blood, or the lumen of the host intestine. A protozoan species might be expected to show specializations associated with (1) the taxonomic group to which it belongs; (2) the habitat (location in the host); and (3) its life pattern, particularly as this relates to its transmission from one host individual to another. These features of specialization should be kept in mind when we come to consider specific examples of protozoan parasites.

Clearly, parasitism has independently evolved in various major groups of protozoans along quite independent lines. In some instances we may find what appear to be ecologically intergrading forms. For example, the dysentery amoeba, *Entamoeba histolytica,* is an obligate parasite. In nature it can reproduce only in a host. However, its congeneric relative, *Entamoeba moshkovskii,* lives quite happily in sewage. Various forms of *Acanthamoeba* seem to be capable of living as parasites on a facultative basis when opportunity favors the establishment of such relationships, but they may also reproduce as free-living independent organisms. On the other hand, all sporozoans are obligate parasites; and many of them live a predominantly intracellular life in the host. Even within a major group, independent origins of parasitism seem to have occurred. The trichomonad flagellates and the hemoflagellates (examples of both will be discussed presently) clearly had different origins among free-living flagellates. Hence one of our determinants of specialization among parasitic protozoans might be elab-

Protozoan Parasites

chapter 2

orated to include consideration of the taxonomic groups of *free-living* organisms from which the particular parasitic forms are most immediately derived.

In this chapter we shall consider some specific examples of parasitic protozoans having independent evolutionary origins and showing qualitatively different adaptations for parasitism.

PLASMODIUM: THE MALARIA ORGANISMS

The protozoans of the genus *Plasmodium* are best known as the parasitic organisms involved in the diseases collectively known as malaria. Three species are primarily responsible for human malaria: these are *P. vivax, P. falciparum,* and *P. malariae.* A fourth species, *P. ovale,* is a more rare human parasite. Malaria has been, and is presently, a serious human disease. It is found mainly in the tropics, although it was formerly of considerable significance in the public health of the United States and Europe. A few years ago the conquest of malaria in America north of Mexico and in some of the other economically advanced countries of the Old World led to enthusiastic support of the concept that malaria could be eradicated on a global basis. Unfortunately eradication has not progressed at the rate anticipated by its proponents, and it is now suspected that eradication does not depend on simple single solutions such as insecticide application or treatment of clinical cases. Except for insular situations, eradication of human malaria may depend on a complex of socioeconomic factors that lead to overall interruption of the life pattern of the *Plasmodium* involved. This interpretation seems inescapable when we objectively examine the eradication of malaria in the United States. Success cannot be attributed to any single control measure. Screens on houses, control of mosquitoes, and treatment of patients all seem to have played a part.

With the waning of domestic malaria problems, the interest of U.S. and European scientists in malaria faltered after World War II, and we recently found ourselves in an embarrassing situation with *falciparum* malaria in South Viet Nam. The sharply delimited geographical areas involved may have saved us from a disaster, but the lesson should not be disregarded. Further, it has become apparent that in large malarious areas in the tropics, human cases of malaria are not always curable with the new synthetic antimalarial drugs: The parasites are sometimes drug-resistant. As a consequence of this more recent experience there has been enhanced interest in malaria during the past few years.

The Life Pattern of *Plasmodium vivax*

The *sporozoites* of this parasite are injected into a human host along with the salivary secretion of an infected anopheline mosquito. The parasite enters fixed cells of the host body, often in the liver, and undergoes at least two cycles of multiple division. As the parasites are liberated from these fixed cells, they invade circulating erythrocytes. Here they undergo growth and nuclear division. This tends to occur synchronously in a large number of erythrocytes, and rounded cells called *merozoites* are formed. The new crop of parasites is liberated and enters fresh erythrocytes where the process is repeated. After a few generations some of these merozoites do not undergo nuclear division in a freshly invaded erythrocyte but develop into large uninucleate organisms. These are *gametocytes,* some of which are male (*microgametocytes*) and some female (*macrogametocytes*). When gametocytes are ingested by a mosquito, the microgametocytes rapidly undergo endomitosis and extrusion of six or eight flagella. A nucleus becomes associated with each flagellum, which then breaks free as a *microgamete.* The macrogametocyte forms a *macrogamete,* and fusion of the two gamete types occurs in the stomach of the mosquito. The resulting zygote is an elongate *ookinete* that penetrates the wall of the stomach and lodges under the outer limiting membrane, forming an *oocyst.* The oocyst undergoes numerous cell divisions, growing in size and forming a bundle of spindle-shaped cells (*sporozoites*). Ten to twenty days after the infective blood meal, the sporozoites break out of the oocyst into the hemocoele of the mosquito. They make their way to the salivary glands in the anterior part of the hemocoele and invade the lumen of these glands. These are infectious for a new vertebrate host and are injected when the mosquito feeds. This pattern is summarized in Fig. 2–1.

Development in a Mosquito

The formation of male gametes in the stomach of a mosquito is an amazing process. In a matter of 8 to 12 minutes, a very complex series of events occurs, resulting in the formation of eight motile gametes. Garnham and his colleagues have worked out the details of this development (Fig. 2–2). The nucleus divides three times endomitotically, forming eight nuclei that are symmetrically arranged around the old nuclear membrane. The centriole also replicates and each product is associated

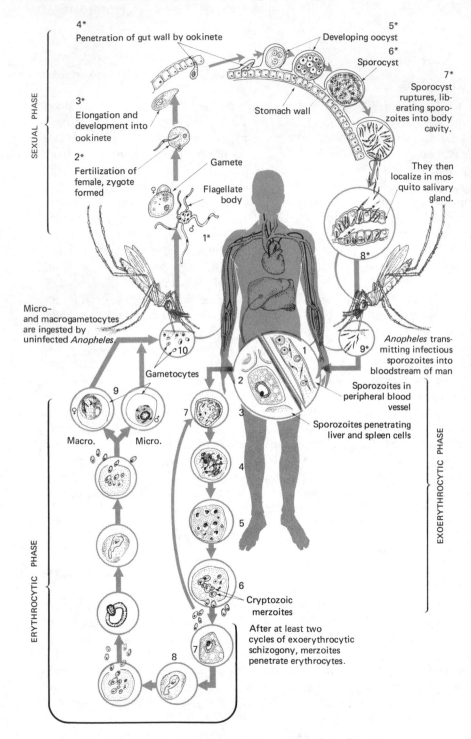

Fig. 2—1. The life pattern of the malaria parasite, *Plasmodium vivax*.

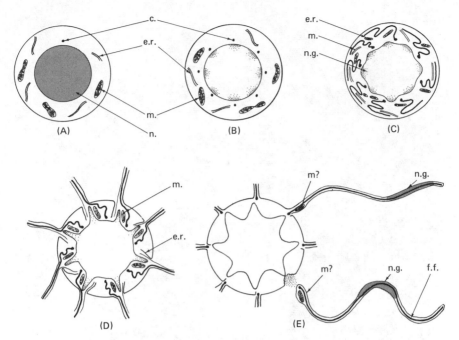

Fig. 2–2. A diagrammatic representation of the formation of male gametes in *Plasmodium*. This occurs in the stomach of the mosquito host. (A) Microgametocyte with a single nucleus and a centriole. (B) Endomitosis and centriole replication. (C) Formation of flagellar fibrils. (D) Association of nuclei and mitochondria near point of emergence of fibril complexes. (E) Fully formed microgametes breaking free. Abbreviations: c., centriole; e.r., endoplasmic reticulum; f.f., flagellar fibrils; m., mitochondrion; n., nucleus of microgametocyte; n.g., nucleus of microgamete. (After Garnham, et al., 1967.)

with a nucleus. The centriole becomes the basal body of the flagellum. The fully developed microgamete of *Plasmodium* is essentially a flagellum, with the usual 9 + 2 fibrils, accompanied by a nucleus.

After fertilization of the female gamete has occurred, the zygote elongates, forming a motile ookinete (Fig. 2–3). Electron microscopy has shown that the ookinete has a pellicle made up of two unit membranes, and that immediately below this pellicle is a row of longitudinal fibrils. These fibrils are thought to function in locomotion. The ookinete also contains several aggregations of what appear to be crystalloids. Since these crystalloids are not found in the preceding gametocyte stage nor in the later oocyst, it has been suggested that they are involved in the formation of the oocyst wall, but they might also be involved in the penetration of the mosquito gut.

The ookinete penetrates the stomach epithelium but goes no further after reaching the basement membrane. Here the parasite rounds up to

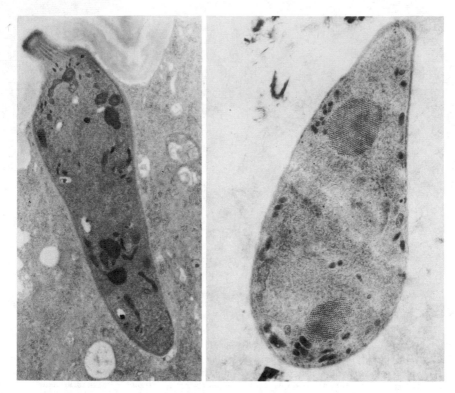

Fig. 2–3. Longitudinal sections through the ookinetes of two malaria parasites. (Left) *Plasmodium berghei yoeli* ($\times$ 9,700). (Right) *P. cynomolgi bastianellii* ($\times$ 10,000). (Courtesy of Professor P. C. C. Garnham.)

form an oocyst in which sporozoites develop. Its localization here is apparently not a nutritional necessity, since Weathersby showed that if gametocytes are injected into the mosquito hemocoele, oocyst development will occur in a variety of locations.

The Feeding of *Plasmodium*

Although it was long recognized that the erythrocytic stages of *Plasmodium* must feed on host cytoplasm and digest hemoglobin (there isn't much else to eat inside a mammalian red cell), the mechanism of feeding was not understood. It was speculated that digestion was extracellular, and there were arguments as to whether the malaria pigment, a degradation product of hemoglobin, was inside the parasite cell or on the surface. Rudzinska and Trager first demonstrated phagotrophy in the rodent malaria organism, *P. berghei*, in 1959. They reported that host cytoplasm is engulfed by invagination of the parasite surface membrane, followed by pinching off of a food vacuole. Similar observations were subsequently reported with several other species of *Plasmodium*. In some cases the vacuoles formed are quite small, and the process seems quite comparable to what has been termed "pinocytosis." In 1966, Aikawa and his colleagues described a somewhat different process in some plasmodia from birds. They reported that the cell had a cytostome, a structure very much like the micropyle that was described by Garnham for other stages of malaria parasites. The cytostome was described as functioning in the ingestion of host cytoplasm. A baglike invagination inside the cytostome pinches off and, with the enclosed host cytoplasm, forms a vacuole (Fig. 2–4). The process is repeated, with the formation of a new invagination inside the cytostome. It has been suggested that both patterns of feeding occur, but the issue is not yet satisfactorily resolved. In spite of this state of affairs, it seems clear that host cytoplasm is ingested, and it is apparently digested in vacuoles that vary in size with the particular parasite species.

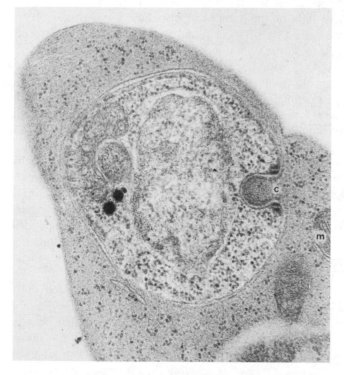

Fig. 2—4. The feeding of the erthrocytic phase of *Plasmodium cathemerium*. Abbreviations: c = cytostome; m = mitochondria of host cell ($\times$ 41,500). (Courtesy of Dr. M. Aikawa and the *Journal of Cell Biology*.)

The characteristic malaria pigment found in the parasites, as well as in host cells, is a hemoglobin degradation product. It is an incompletely digested heme protein. Earlier reports that *Plasmodium* cleaves hemoglobin into a heme and a globin moiety do not seem to be consistent with the type of product represented by malaria pigment.

As will be discussed below, there is also evidence that malaria parasites feed on low molecular weight metabolites that may enter the red cell from the host plasma or may be synthesized by the host cell.

Nutrition of *Plasmodium*

The widespread occurrence of malaria in military personnel during World War II stimulated research on the nutrition of malaria parasites. Much of this work was grounded in the hope that an understanding of nutrition would lead to a rationally devised chemotherapy for malaria. Although the applications of these nutritional studies to chemotherapy have not been dramatic, much new information has been obtained, some of which sheds light on the nature of the malaria parasitism.

In the early 1940s Trager found that the survival in vitro of the avian parasite, *P. lophurae,* was markedly benefited by the addition of pantothenic acid to a complex medium containing infected bird erythrocytes. Glucose was also required. In this same period Geiman and his colleagues cultivated mammalian erythrocytes infected with the simian parasite *Plasmodium knowlesi.* It was shown that in addition to glucose the organism required *p*-aminobenzoic acid and methionine. The culture medium contained plasma, and it proved impossible to define other requirements. Further, it was shown that *P. knowlesi* is sensitive to sulfonamide drugs and that the inhibition produced by sulfonamides was reversed by *p*-aminobenzoic acid. Extrapolating from bacteria, this was consistent with the view that the parasite required *p*-aminobenzoic acid for the synthesis of folic acid. However, the presence of the host cells allowed only tentative conclusions as to the role of the above mentioned compounds in the nutrition of *Plasmodium.* As we will see, these tentative conclusions were not entirely correct.

Over a period of several years Trager has attempted the culture of *Plasmodium* in cell-free media, with the recognition that direct nutritional requirements could only be determined with the isolated parasite. These media have been quite complex, but it was discovered that the addition of pantothenic acid or *p*-aminobenzoic acid did not promote survival of the parasite. The addition of Coenzyme A, the pantothenic acid-containing cofactor, promoted survival. Thus it would appear that the apparent pantothenic acid requirement is indeed a requirement for Coenzyme A, a substance normally found only inside cells. Similarly it was found that the survival of *Plasmodium* was promoted by adding folinic acid rather than *p*-aminobenzoic acid. Again this is a cofactor that is synthesized by the host cell.

It is of interest to comment on the effects of those compounds that stimulate *Plasmodium* growth in the intact parasite-host cell system. The enhancement of parasite growth on the addition of pantothenic acid might imply that the normal amounts of this material available for the synthesis of Coenzyme A do not satisfy the requirements of the two-membered system. Might we also infer that the parasite redirects the metabolism of the host cell so that larger than usual amounts of Coenzyme A are synthesized? The same question could be asked about folic acid metabolism. The folic acid coenzyme nutrition of *Plasmodium* has been difficult to decipher. In the intact parasite-host cell system, *p*-aminobenzoic acid is required, and it is probable that the parasite carries out the reaction

$$\text{Pteridine} + p\text{-aminobenzoic acid} \rightarrow \text{folic acid.}$$

However, the parasite, freed from the erythrocyte, is not stimulated by *p*-aminobenzoic acid, nor is it stimulated by folic acid. Its growth is stimulated by folinic acid (N^5-formyltetrahydrofolic acid). Further, there is evidence that *Plasmodium* can convert dihydrofolic acid to an active coenzyme form but cannot make such a conversion of folic acid. Thus there appears to be a defect in the conversion of folic acid to dihydrofolic acid.

It was also found that the survival of *Plasmodium* was enhanced by the addition of adenosine triphosphate, NAD, malic acid, and pyruvic acid. It is of considerable interest that these organisms may be permeable to highly polar substances, such as adenosine triphosphate, and suggests that *Plasmodium* may be an energy parasite. Thus, although it has been shown that *Plasmodium* has the kinases for the synthesis of adenosine triphosphate from adenosine diphosphate in the glycolytic sequence, it is significant that erythrocytes infected with *P. lophurae* have a lowered adenosine triphosphate content and that *P. falciparum* develops more rapidly in cells having a high adenosine triphosphate content than in cells with a lower level of this energy-rich compound. Also adenosine diphosphate does not substitute for adenosine triphosphate in the culture of free *P. lophurae*. The NAD content of cells infected with *Plasmodium* is considerably higher than that of uninfected cells. This has been interpreted as a case in which NAD synthesis by the host cell has been redirected by the parasite. Also, it may be significant that *Plasmodium* cannot synthesize adenine, a constituent of both adenosine triphosphate and NAD.

The requirement for an external source of methionine by the parasite-host cell system is an interesting one. It has been shown that sulfur-labeled methionine in hemoglobin is utilized by *Plasmodium* in protein synthesis. However, primate hemoglobins have a relatively low methionine content. It seems likely that the additional methionine requirement is related to the special role of methionine in methylation reactions involving formation of S-adenosylmethionine; thus,

$$\text{Methionine} + \text{adenosine triphosphate} \rightarrow \text{S-adenosylmethionine.}$$

More recently it has been shown that parasite growth in the *P. knowlesi*-red cell system of monkeys requires the addition of the amino acid isoleucine. In the absence of added isoleucine no parasite growth occurs. Primate hemoglobins are devoid of isoleucine. Further, with isoleucine in the culture medium the isoleucine antagonist methyl-0-threonine inhibits growth of *P. knowlesi*. This inhibition is reversed by the addition of more isoleucine.

Although the nutritional requirements of the malaria parasites are far from completely known, the foregoing suggests a sufficient simplification of synthetic capacities to explain the dependence of these organisms on the host cell. Not yet examined are such parameters as the presence or absence of ion gradient-coupled membrane transport in these organisms that may represent even more fundamental forms of dependence on the host.

Energy Metabolism in *Plasmodium*

Malaria parasites do not seem to accumulate storage polysaccharides and are presumably dependent on a constant supply of oxidizable carbon compounds from host sources.

Erythrocytes harboring malaria parasites metabolize sugar and take up oxygen at considerably higher rates than uninfected cells and, under anaerobiasis, produce lactic acid at very high rates. Studies on cell-free extracts of *Plasmodium gallinaceum,* an avian parasite, showed the presence of several Embden-Meyerhof sequence enzymes in the parasites. Subsequently, similar observations have been made on other species of *Plasmodium*. The enzymes of the Krebs cycle have also been demonstrated in species from mammalian and avian hosts and seem to function in the oxidation of carbohydrates. However, fermentation proceeds at a high rate, and even under aerobic conditions at least four to six moles of glucose are degraded to lactate for each one oxidized. Further, intermediates of the Krebs cycle may "leak" from the organism. Thus from the standpoint of energy production the Krebs cycle does not appear to play the major role assigned to it in some other animal organisms.

One difficulty in studying the metabolism of the isolated *Plasmodium* is the removal of the parasites without damage from infected erythrocytes.

Further, the suspending medium may be of great importance in determining overall metabolic rate and the relative operation of particular pathways in metabolism. The ionic composition of many suspending media used in the past bears little resemblance to the ionic composition of the environment furnished by the red cell. Since *Plasmodium* cells appear to be "leaky," the character of the medium may be more important than is the case with some other organisms.

Differences between analogous enzymes of *Plasmodium* and the host cells have been demonstrated. Sherman showed that the lactic dehydrogenases of *P. lophurae* and *P. berghei* can be differentiated from host enzymes by the net charge on the proteins, affinities for NAD analogues, substrate inhibition, and effects of pH on enzyme activity. Similarly the malic dehydrogenase of *P. lophurae* differs from that of the host in pH optimum, Michaelis' constant, and affinity for NAD analogues. The sensitivity of phosphofructokinase of *Plasmodium* to the antimalarial drug mepacrine is indicative of a significant difference between this enzyme and the analogous one in the host.

Malaria parasites, living in erythrocytes that are normally low in glucose-6-phosphate dehydrogenase and 6-phosphogluconic acid dehydrogenase, cause an increase in the level of these enzymes in infected cells. It was suggested that this might make available increased quantities of reduced coenzymes for biosynthetic processes. However, there is no real evidence that the parasite is incapable of reducing coenzymes, and the basically fermentative character of energy metabolism should yield an adequate reducing capacity.

Malaria Transmission

A further word must be said concerning the relationships of *Plasmodium* to the invertebrate hosts. The mammalian species are transmitted only by mosquitoes of the genus *Anopheles*. Avian forms may utilize mosquitoes of the genera *Culex, Aedes,* and others. There is an enormous literature relating to the distribution, taxonomy, and ecology of the mosquitoes, particularly as they relate to human disease. Certain general statements are possible. The species of *Anopheles* vary greatly in their capacity to support the development of a given *Plasmodium* species. Further, within a particular mosquito species there is variation in the degree to which various strains are susceptible to a single strain of *Plasmodium*. It has been shown on several occasions that a local mosquito strain, serving as a satisfactory host and as an efficient transmitter to human hosts for a local strain of *Plasmodium,* may not be a very satisfactory host for some other "foreign" strain of *Plasmodium*. There is some evidence that the rate of formation of the peritrophic membrane and the rate of secretion of digestive enzymes may be involved. Although little is known of the physiological bases for these adaptations, there is considerable evidence that there is a genetic background.

Transmission to the vertebrate host will also depend on whether or not the mosquito feeds on the host. Some *Anopheles* are not significant vectors of human malaria because they prefer other animals than man for food. What has been termed critical density is important; this refers to the average number of *Anopheles* bites per person per day. If this number falls below a certain level, there is a progressive decrease in malaria. Critical density will be determined by the number of mosquitoes, their predilection for humans as food, their length of life, their frequency of feeding, the proportion of mosquitoes becoming infected, and a number of other parameters.

Plasmodium and "Abnormal" Red Cells

It has been known for several years that there is a relationship between the occurrence of sickle cell trait in humans and the severity of malaria infections. Individuals who are homozygous for sickle cell trait suffer from an anemia, and death usually occurs at an early age. It has been shown that the trait involves an abnormal hemoglobin, (S), which differs sharply in its properties from normal hemoglobin, (A). Hemoglobin S differs from Hemoglobin A in replacement of a glutamic acid residue by valine. The loss in charge causes a marked reduction in solubility and an increase in viscosity.

Persons who are heterozygous have both S and A and appear to be quite healthy. Such persons harbor fewer malaria parasites and have a lower mortality from malaria than those who are homozygous for A. This is particularly significant in nonimmune children and is of selective advantage in human populations living in regions where *falciparum* malaria is endemic. Theoretically, in these populations one-fourth of the children may die from sickle cell anemia, one-fourth may have severe, perhaps fatal, malaria, and one-half (heterozygotes) may survive to reach reproductive age. Clearly, if malaria is eradicated, or markedly reduced, there would be a concomitant decrease in the biological advantage accruing from maintenance of sickle cell trait in these populations.

The physiological basis for relative insusceptibility to malaria conferred by sickle cell trait is not well understood. It has been suggested that (1) parasites may not enter cells containing S, (2) the viscosity of S interferes with ingestion of host cytoplasm by the parasite, and (3) modification of red cells containing S (and parasites) renders them more subject to destruction by the lymphoid-macrophage system of the host. Further study may reveal which of these or what other possibilities explain the immunity conferred by sickle cell trait.

There have been reports that thalassemia trait, an aggregation of human hemoglobin abnormalities, may confer insusceptibility to malaria. Similarly, a glucose-6-phosphate dehydrogenase genetic deficiency of the host has been purported to furnish some immunity to *falciparum* malaria. However, there seems to be some conflict among different researchers concerning the connections of these factors to malaria.

HEMOFLAGELLATES: THE TRYPANOSOMES

The trypanosome protozoans are among the most graceful of animals. The thrill of observing the undulations of these organisms as they move among the red cells in a wet smear of mammalian blood is certainly at least equivalent to observing a highly colored tropical bird in its native haunts. Paul Ehrlich's preoccupation with these organisms in his early search for specific chemotherapeutic drugs is easily understood when one considers the fascinating characteristics of the organisms. However, their appearance is deceptive. The trypanosomes have rendered huge areas of the most fertile parts of Africa quite unsuitable for habitation by man or his domestic animals. More than 25 percent of the arable land of Africa is unfit for agricultural purposes because of the mortality produced by trypanosome diseases (Fig. 2–5). In South and Central America, Chagas' disease, a trypanosome-produced malady of man, may be on the increase. It is a disease of the poor in the Americas, and the vast tropical ghetto developing in that part of the world may produce an enhanced incidence. At this writing there is no cure for Chagas' disease, and the victims die a slow and singularly unpleasant death. Whereas healthy U.S. citizens worry about dying from chronic diseases brought on by overeating and other forms of self-indulgence, millions of people suffer from an incurable disease whose origin is well understood and which could be controlled by decent housing. If no other justification existed for studying parasitism, the humanitarian impulse should constitute a pressing reason for studying trypanosomes and trypanosomiasis. Intelligent self-interest is of equal significance.

In order to understand the terms applied to various forms of hemoflagellate protozoans, it seems desirable to outline the terminology in a highly abbreviated fashion. The morphological types occurring in the genera *Trypanosoma* and *Leishmania* are shown diagrammatically in Fig. 2–6. For a more detailed discussion of cell types in various genera of hemoflagellates, consult Hoare (1967).

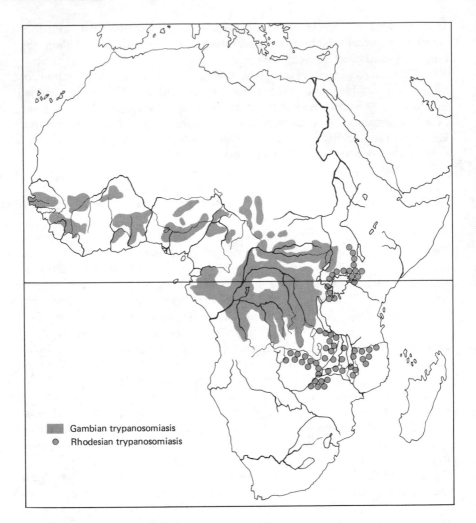

Fig. 2—5. The geographical distribution of African trypanosomes causing human disease.

The tsetse fly-borne African trypanosomes of the *brucei* subgroup have been studied for some time and will be used as major examples for discussion.

The *brucei* Subgroup of Trypanosomes

Trypanosoma brucei is a widespread parasite of African mammals, except for man and baboons. In domestic animals it is unusually virulent and produces a high mortality. There seems to be little doubt that *T.*

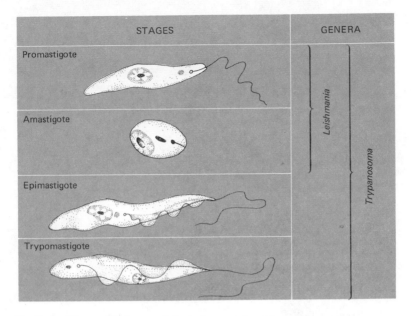

Fig. 2—6. Diagram of the morphological stages found in the genera *Trypanosoma* and *Leishmania.*

brucei is the parent form from which two species that infect man have been derived. These two forms are *T. gambiense* and *T. rhodesiense,* and the three species collectively are referred to as the *brucei* subgroup. In the blood of the vertebrate host these trypanosomes, unlike some other species, are polymorphic, some individuals being slender with a free flagellum, some being short and stumpy without a free flagellum, and some intermediate in morphology (Fig. 2–7). It is impossible to distinguish the three species morphologically, although they show differences in geographical and host distribution and in the characteristics of the diseases produced. All undergo cyclic development in tsetse flies of the genus *Glossina* and are transmitted to the vertebrate host by the biting of these flies.

Wild animals may serve as reservoirs for *T. gambiense* and *T. rhodesiense,* particularly for the latter. There are important differences in the behavior of the two parasites in human hosts. *T. rhodesiense* infections typically appear in thinly populated areas where game animals are common, whereas *T. gambiense* infections occur in more populated areas. *T. gambiense* produces a slowly developing chronic disease that may go on for several years before the so-called sleeping sickness symptoms appear. These symptoms herald the invasion of the central nervous system by the organisms and are manifested as progressive physical and mental depression. *T. rhodesiense* produces a similar disease, but it is more acute and runs its course in a few months. *T. rhodesiense* tends to occur in sporadic

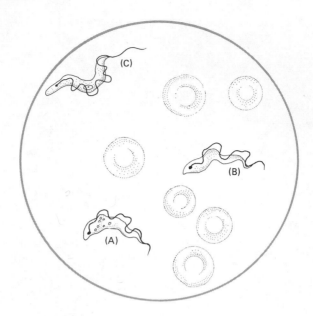

Fig. 2–7. The forms of *Trypanosoma gambiense* found in the blood of the vertebrate host: (A) stumpy form; (B) intermediate form; (C) slender form.

epidemics, whereas *T. gambiense* smolders and in the long run produces higher human mortality.

When all factors are considered, however, *T. brucei* infections in domestic animals are probably of greater consequence to human welfare than the infection of men by *T. gambiense* and *T. rhodesiense*. Enormous areas of fertile African land are virtually uninhabitable by men simply because domestic mammals cannot survive. This is a great disadvantage for a protein-poor continent.

The Life Pattern of *T. gambiense*

The trypanosomes, or *trypomastigotes,* in the blood of the vertebrate host are ingested by a tsetse fly, commonly *Glossina palpalis,* along with a blood meal. The parasites multiply in the midgut of the fly, and after a period of some days they move forward to the salivary glands where they undergo further multiplication as *epimastigote* forms. Eventually trypomastigotes are again formed and are infective for a vertebrate host. When the fly once more bites a man, trypanosomes are transmitted to the new vertebrate host. In the vertebrate, multiplication occurs in the extracellular body fluids. Sexual reproduction has never been convincingly demonstrated. The general pattern is shown in Fig. 2–8.

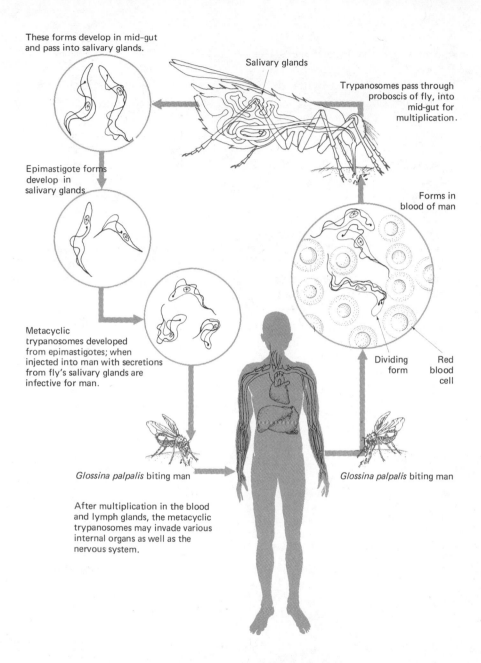

These forms develop in mid-gut and pass into salivary glands.

Salivary glands

Trypanosomes pass through proboscis of fly, into mid-gut for multiplication.

Epimastigote forms develop in salivary glands

Forms in blood of man

Metacyclic trypanosomes developed from epimastigotes; when injected into man with secretions from fly's salivary glands are infective for man.

Dividing form

Red blood cell

Glossina palpalis biting man

Glossina palpalis biting man

After multiplication in the blood and lymph glands, the metacyclic trypanosomes may invade various internal organs as well as the nervous system.

Fig. 2—8. The life pattern of *Trypanosoma gambiense.*

Development in the Invertebrate Host

After ingestion by the fly host, *T. gambiense* multiplies in the middle intestine for 10 to 15 days. However, not all flies become infected, and the factors that determine susceptibility appear to be complex. The temperature at which the pupa of *Glossina* develops seems to affect susceptibility of the adult fly. The age of the fly at the time it feeds on infected blood is also important. Wijers found that flies that fed on an infected vertebrate on the first, second, or subsequent days after emergence from the pupal stage showed infection rates of 7.6, 1.1, and 0 percent, respectively. It was hypothesized that this is related to the development of the peritrophic membrane in the flies.

After 10 to 15 days the trypomastigotes become very elongate and migrate forward to the proventriculus of the arthropod. After several more days they move forward to the salivary glands, becoming attached to the wall of the gland and multiplying further as epimastigote forms (Fig. 2–9). Finally, 20 to 30 days after being taken in by the fly, the so-called metacyclic trypanosomes, trypomastigote forms, appear in the lumen of the salivary glands and are infectious to a vertebrate. During feeding on a vertebrate, *Glossina* intermittently injects salivary secretion and thus transmits the protozoan to a vertebrate host.

Fig. 2–9. Epimastigote forms of *Trypanosoma gambiense* found in the salivary glands of the fly host. (After Wenyon.)

A considerable effort has been made to determine the stimuli for the developmental sequence of the trypanosomes in the fly host. During culture in vitro, the addition of trehalose, arabinose, or inositol has been reported to stimulate development of forms infective to vertebrates. However, results have been conflicting and further work is needed to clarify the confusion. We shall discuss the differences in the biochemistry of the invertebrate and vertebrate forms of the trypanosomes presently.

Cell Division in Hemoflagellates

The division of cells has been studied most extensively in culture forms. The first event noted is a lateral movement of the kinetoplast, so that it lies very close to the cell membrane. The basal granule of the flagellum then divides, with the original flagellum remaining attached to one of the granules and a new flagellum originating from the other. The kinetoplast then divides, and while this is occurring, a projection appears on the cell. The angle between this projection and the old cell delineates the area of separation of the daughter cells to be formed. The nucleus undergoes mitosis, and the cell now contains two complete sets of organelles. As the new cells become more separate, they come to lie in opposite directions and eventually separate. These events are diagrammed in Fig. 2–10.

Development of *T. gambiense* in the Vertebrate

The "metacyclic" trypomastigotes transmitted by the bite of an infected tsetse fly are very similar to some of the forms later found in the blood of the vertebrate. There is considerable evidence that in some vertebrate hosts a considerable trypanosome multiplication occurs in the neighborhood of the inoculation site before the organisms appear in the blood. There is, however, no evidence that the organisms go through any required cycle of development during the early period in the vertebrate host. It is known that the proportions of slender and stumpy forms (Fig. 2–7) change as the infection progresses. During the period of increasing parasitemia, the long forms predominate; later the stumpy forms increase in relative numbers. This may be significant in terms of infection of the arthropod host.

Energy Metabolism of the *brucei* Subgroup
of Trypanosomes

The trypanosomes are aerobic organisms. Motility ceases and survival is curtailed in the absence of oxygen. The *brucei* subgroup shows very high rates of carbohydrate metabolism, and it is markedly dependent on the presence of glucose in the suspending medium. The subgroup may

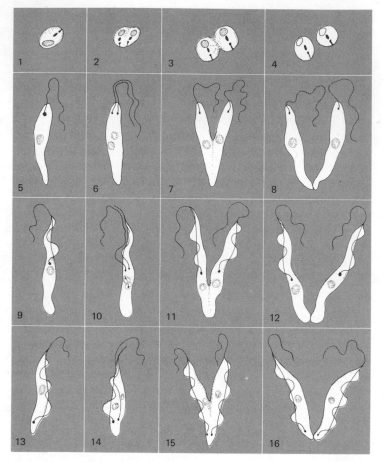

Fig. 2—10. Diagrammatic representation of cell division in various trypano-somatid types: 1—4, division of amastigote; 5—8, division of promastigote; 9—12, division of epimastigote; 13—16, division of trypomastigote. (Modified from Wenyon, 1926.)

consume in an hour an amount of sugar corresponding to its cell weight. This utter dependence on exogenous glucose is related to the fact that the trypanosomes do not store significant amounts of carbohydrate. There are gross differences in the metabolism of the blood-stream forms and the culture forms* of *T. gambiense* and *T. rhodesiense,* the blood forms using as much as ten times more glucose per unit time. The blood forms produce negligible amounts of carbon dioxide in the oxidation of glucose, whereas the culture forms produce carbon dioxide essentially equivalent in

---

* The culture forms are morphologically and presumably physiologically similar to the forms found in the insect host.

amount to the oxygen consumed. The end products of metabolism also differ in that the blood forms excrete pyruvic acid in the presence of oxygen and pyruvic acid plus glycerol under anaerobic conditions, whereas the culture forms produce acetic and succinic as well as pyruvic and lactic acids. Further, the respiration of the blood forms is quite insensitive to cyanide or azide, whereas that of the culture forms is inhibited by these compounds. These observations indicate that the morphological differences between the blood and insect phases are accompanied by profound differences in energy metabolism.

The foregoing suggests that (1) electron transport mechanisms differ in the blood and culture forms, (2) pathways for the complete oxidation of carbohydrate are found in the culture forms but not in the blood forms, and (3) direct morphological manifestations of these differences should be demonstrable.

The cyanide-sensitive culture forms of *T. gambiense* and *T. rhodesiense* contain hemoproteins yielding absorption spectra that appear analogous to those of the cytochrome components found in yeasts and mammals. However, preparations of culture forms do not oxidize reduced mammalian cytochrome *c*. As might be expected, the blood forms of these trypanosomes contain negligible amounts of these hemoproteins. The data suggest that some kind of cytochrome system functions in the culture forms. On the other hand, Grant and his colleagues have shown that electron transport in the blood forms of the *brucei* trypanosomes involves an aerobic glycerophosphate dehydrogenase, and they have suggested that a substrate specific peroxidase results in reoxidation of reduced nicotinamide adenine dinucleotide (NADH). Thus,

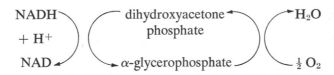

The overall operation of this system in the degradation of glucose to pyruvic acid is thought to follow the pattern shown in Fig. 2–11. Clearly this pyruvic acid fermentation would require large quantities of oxygen, and it is not surprising that the blood forms of *T. gambiense* and *T. rhodesiense* are rapidly immobilized under anaerobic conditions.

A further difference between the blood forms and the culture forms is the lack of a Krebs cycle in the blood forms. This is, of course, related to the mechanism utilized for the reoxidation of NADH discussed above. The culture forms have a Krebs cycle, and about half of the glucose metabolized is oxidized to carbon dioxide and water. Estimates of the equivalent numbers of phosphate bonds that may be generated in the metabolic patterns of the blood and culture forms suggest that the culture forms should

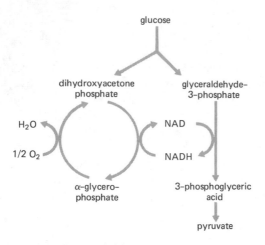

Fig. 2–11. An abbreviated scheme for the degradation of glucose in the bloodstream forms of the *brucei* trypanosomes.

consume about one-tenth as much glucose and about half as much oxygen. This is reasonably close to what has actually been observed. Another subgroup of trypanosomes, including *T. evansi, T. equinum,* and *T. equiperdum,* does not undergo cyclic development in an intermediate host and completely lacks a Krebs cycle.

The production of succinate by the culture forms of the *brucei* trypanosomes may be independent of oxygen consumption and may involve carbon dioxide fixation at the pyruvic acid or phosphoenolpyruvic acid stage. Under anaerobic conditions, there is an increase in succinate production by the culture forms.

Metabolic Polymorphism

As indicated above, there is some polymorphism in the blood forms of these trypanosomes. It was recognized many years ago that if these trypanosomes are transferred from vertebrate to vertebrate by syringe, this polymorphism may virtually disappear, the short stumpy forms (Fig. 2–7) no longer being evident. At the same time the trypanosomes completely lose the capacity to infect arthropods. Wijers and Willett examined flies, after feeding them on a host infected with *T. gambiense,* and concluded that only the stumpy forms were infectious to the arthropod host. In 1964, Balis examined the effects of α-ketoglutaric acid, a Krebs cycle intermediate, on washed cells of *T. brucei* removed from the blood of the vertebrate host. He found that the motility of the short stumpy forms was maintained in the presence of this compound, whereas the slender forms became nonmotile, as they typically do in the absence of an energy source. This suggested that the stumpy forms have a *functional* Krebs cycle. Vickerman extended this work and showed that the intermediate and stumpy forms of *T. brucei* have NADH-tetrazolium reductase activity in the mitochondria. Of great interest is the indication that the switch to the respiratory pattern of the arthropod phase of *T. brucei* actually occurs, at least in part, while the organism is in the blood of the vertebrate. This implies that the "appearance" of a Krebs cycle, and presumably the appearance of some modification of the electron transport system, do not necessarily occur as a response to the environment furnished by the arthropod.

What other observed morphological changes can be related to the changes in the metabolic pattern? In addition to the gross changes from the slender to the intermediate to the stumpy forms, significant changes occur in the single mitochondrion. This contains DNA, but in the slender bloodstream trypomastigotes there are no cristal elaborations of the mitochondrial inner membrane. In culture forms, developing extensions of the mitochondrion have numerous cristae mitochondriales. Vickerman showed that in the transition from slender to stumpy forms of *T. brucei* there was an increase in the diameter of the tubular kinetoplast extensions and an accompanying appearance of cristae.

It appears that in the trypomastigote the single mitochondrion includes the kinetoplast, and the morphogenesis and maintenance of this organelle are probably controlled by kinetoplast DNA. It has been shown that the *brucei* subgroup of trypanosomes can be rendered dyskinetoplastic by treatment with acriflavine, rosaniline, or pararosaniline. Such forms lose the kinetoplast DNA, although the mitochondrial membranes do not disappear. Concomitantly, the trypanosomes lose the capacity to infect the arthropod host or to grow in culture. Reichenow suggested some years ago that the kinetoplast (= mitochondrial DNA) is a metabolic organelle that is not required for the life of African trypanosomes in the blood of the vertebrate host but is essential for survival in the insect. In general this has been borne out by later studies, although it may not be true for all trypanosomes.

Comparisons have been made between the metabolic transformations in trypanosomes and in yeasts, the latter changing from an anaerobic fermentation to the aerobic oxidation of glucose, with an accompanying modification in electron transport systems. In yeasts, oxygen appears to constitute the stimulus, whereas in trypanosomes, oxygen is an unlikely candidate for such a role. Although it has been suggested that the stimulus for trypanosome metabolic transformation may be an immune response of the host, further research is clearly required. No information is available on the biochemical changes actually occurring during cyclic development in the insect, although it has been assumed that the forms appearing in cultures are physiologically, as well as morphologically, equivalent to some of the forms appearing in the insect phase. In the *brucei* trypanosomes, the epimastigote insect form does not usually appear in cultures.

Cultivation of *brucei* Trypanosomes

Although these organisms can be grown on several complex media containing blood or in tissue culture with mammalian or insect cells, virtually nothing is known of their nutritional requirements. There is suggestive evidence that a considerable period of development in culture may be required before the organisms become infective for the vertebrate. If this is indeed true, the study of the physiological cyclic changes in culture might shed further light on what occurs in the insect host.

Drug Resistance

It has been known for many years that microorganisms may become resistant to chemotherapeutic agents. This was first demonstrated by Paul Ehrlich with *Trypanosoma brucei*. He found that when mice infected with *T. brucei* were treated with the dye parafuchsin, some animals were cured. In others, however, the trypanosomes reappeared after a few days. When these recurring infections were treated with dye, there was no therapeutic effect and the animals died. If such parafuchsin-resistant trypanosomes were injected into fresh mice, they produced infections that were unaffected by parafuchsin.

Subsequently, trypanosomes were found to develop resistance to a variety of agents, including trypan blue, a number of organic arsenical compounds, derivatives of phenylstibonic acid, several diamidines, and antrycide, a compound with quaternized quinaldine and pyrimidine rings. Resistance may be quite stable. Fulton and Grant reported that a strain of *T. rhodesiense* had maintained its resistance to the arsenical drug atoxyl for 24 years after its last exposure to the drug. Resistance may develop quite rapidly in the *brucei* trypanosomes. In one area of Congo the first tryparsamide-resistant cases of human trypanosomiasis were observed in 1929, but by 1932, 50 percent of the cases treated were resistant to the drug.

Little is known of the mechanisms of drug resistance in trypanosomes, but in 1960, Inoki and Matsushiro reported that susceptible trypanosomes could be rendered drug-resistant by treatment with DNA from resistant strains. This would deserve further investigation.

Other Biological Considerations

A further word may be said concerning the problem of recognizing species of trypanosomes. The *brucei*-like forms furnish an excellent example of the difficulties. As we stated earlier, the three species concerned are morphologically identical. Their differences are "biological" and are shown in their relationships to hosts. All infect mammals, but only *T. gambiense* and *T. rhodesiense* infect man. These two forms differ in the type of human disease appearing. To complicate matters further, it has been shown recently that in the Zambezi basin and the Botswana there is a third type of human trypanosome, morphologically identical with *T. gambiense* and *T. rhodesiense,* that produces a human disease like *T. gambiense* but behaves like *T. rhodesiense* in laboratory animals. The fly vectors of *T. rhodesiense* and *T. gambiense* also differ in habitat and in food preferences. *Glossina morsitans* and its relatives, hosts for *T. rhodesiense,* live in savannah country, and the flies feed on game animals, particularly antelopes. Man, apparently, intrudes on a fly-game animal cycle of *T. rhodesiense* and becomes infected. In human outbreaks, man → tsetse → man transmission may occur. On the other hand, *Glossina palpalis* and its relatives, vectors of *T. gambiense,* live among trees near rivers. Their main food is obtained from crocodiles that do not serve as hosts for *T. gambiense.* Thus it appears that the disease caused by *T. gambiense* mainly involves man and *Glossina,* although pigs and goats, living close to man, may occasionally serve as reservoir hosts.

Unfortunately these biological characteristics are not totally reliable. In the laboratory, *G. morsitans* and *G. palpalis* can be infected with any of the *brucei*-type trypanosomes. Further, Gambian sleeping sickness is sometimes acute, like the Rhodesian form, whereas Rhodesian sleeping sickness occasionally runs a chronic course as do *T. gambiense* infections. As a matter of fact these forms should be regarded as races of *T. brucei,* and their separation as species has only been justified on the basis of convenience. The discovery of the Zambezian intergrade (mentioned above) raises the question as to whether the specific separation of these forms may not have obscured the true status of the races of *T. brucei.*

In describing the complications of the biology of the *brucei* group of African trypanosomes, we implied, but did not explicitly state, that the races (or so-called species) of these forms in man might be of recent origin. At the crudest level we know that some of these trypanosomiases have moved about Africa with human intervention. Speciation has probably followed this pattern. There is reasonably good circumstantial evidence that African sleeping sickness in humans was introduced into Uganda from the west coast of Africa by one of the porters accompanying the explorer Stanley in 1887. During subsequent years it had catastrophic effect on human populations. Between 1901 and 1905, more than 200,000 people out of a population of 300,000 died in eastern Africa, in what was then Uganda. This included the startlingly beautiful country around Lakes Victoria, Albert, Edward, and George. There is reason to believe that until human traffic increased, there was essentially a single form of human African sleeping sickness, more or less of the type that we call Gambian sleeping sickness.

Human movement has also modified the distribution and the nature of trypanosomiasis among the domestic animals of man. During the past 50 years, trypanosomiasis of domestic animals has spread remarkably in Africa, entering areas in which the native population was very dependent on domestic animals for sustaining life. By 1946, at least a half of Uganda Protectorate was uninhabitable by cattle. In a single district, Buruli, the population of domestic cattle dropped from 13,500 to 150 during the years 1940–1945. The spread of these diseases of domestic animals has of course intensified the problems of overcrowding in the uninfected areas and of the gross overuse of arable land—conditions that have contributed more than the politicians will acknowledge to the instability of the area bordering Congo and Kenya. Political and social stability in this area of Africa can be attained only if the trypanosomiases of man and animals are brought under control. The Zambezi strain of human trypanosomiasis (p. 33) also probably arose from the shifts in human population in this region, and it seems very likely that other races of *brucei* trypanosomes will be discovered in man.

HEMOFLAGELLATES: THE LEISHMANIAS

These interesting hemoflagellates of the genus *Leishmania* are parasites of mammals and lizards and utilize small biting sand flies (*Phlebotomus*) as intermediate hosts. *Leishmania* infections of man are scattered over wide areas of the globe from China across Asia to India, Iran and Afghanistan, the Middle and Near East, the Mediterranean basin, into Sudan and Ethiopia and East and West Africa. In the New World they extend from Mexico to the northern part of Argentina. There is no continuity in the distribution over these wide areas, nor is a single disease involved. The distribution is determined by the kinds of other mammals that can serve as hosts and by the distribution of the sand fly intermediate hosts. The diseases attributed to *Leishmania* range from a mildly inconvenient lesion of the skin, known as Oriental sore, to a serious disease involving the liver and spleen, known as kala azar. In South America one form of leishmaniasis (mucocutaneous) extensively involves the mucosa of the mouth, nose, and pharynx, as well as the skin, but does not affect the viscera.

The Life Pattern of *Leishmania*

The amastigote forms of *Leishmania* are ingested by sand flies when they feed on the blood and tissue fluids of a vertebrate. Compared with trypanosomes, development of *Leishmania* in the sand fly seems deceptively simple, involving the transformation of the amastigote forms to *promastigotes*. These multiply in the digestive tract of the fly and may be injected into a new vertebrate host when the fly feeds again. Experimentally, it is necessary for sand flies to feed on fruit juices before they can be infected with *Leishmania*. In the vertebrate the promastigotes enter host cells and round up and lose the flagella, forming the so-called leishman bodies, or amastigote forms, that are the characteristic forms found in the vertebrate. The events of the life pattern are shown in Fig. 2–12.

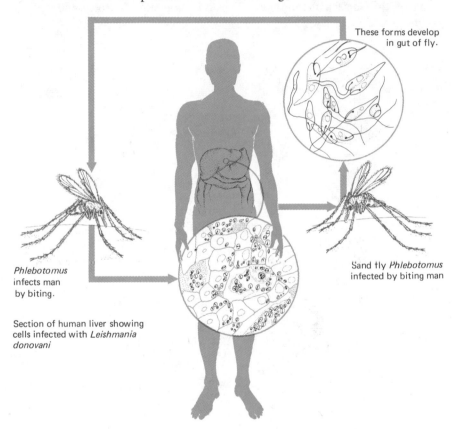

These forms develop in gut of fly.

Phlebotomus infects man by biting.

Sand fly *Phlebotomus* infected by biting man

Section of human liver showing cells infected with *Leishmania donovani*

Fig. 2–12. The life pattern of *Leishmania donovani*.

Development of the Promastigote

As already discussed in connection with the trypanosomes, there are remarkable changes in the morphology of the hemoflagellates during the cyclic development in invertebrate and vertebrate hosts. In the case of

Leishmania there are only two morphologically different stages in the life cycle. During the transition of the amastigote to the promastigote stage in *L. donovani,* there are dramatic changes in the mitochondrion. This organelle is small and undifferentiated in the amastigote stage and contains a central clump of DNA. As development to the promastigote proceeds, the mitochondrion enlarges, without increase in DNA, and develops branches and numerous cristae. This is accompanied by an increase in cytochrome components. Accompanying the changes in the mitochondrion, there are changes in the nucleoli, flagellar growth, increase in cell size, and modifications of the surface membrane. The cell elongates to form the typical promastigote (Fig. 2–13).

Fig. 2—13. The forms of *Leishmania donovani* found in culture or in the sand fly host.

Treatment of the mastigote stages with acriflavine results in what are called dyskinetoplastic cells. Such cells are no longer capable of the morphological transformation described above and appear to have lost the kinetoplast DNA. If the treated cells are in the amastigote stage, they may not be capable of continuing growth in the vertebrate host. Dyskinetoplastic promastigotes do not appear in culture. Trager and Rudzinska found that the effects of acriflavine on *L. tarentolae* are antagonized by riboflavin. The DNA of *L. enrietti* has been studied in the analytical centrifuge. The mitochondrial DNA has a genuine-cytosine content of 36 percent, whereas that of nuclear DNA is 57 percent.

Energy Metabolism

Our knowledge of the metabolism of *Leishmania* is quite incomplete. The promastigote stages may be able to utilize a broader range of compounds as energy sources than the culture forms of some of the trypanosomes. *L. donovani* culture forms have a sucrase, for example, and some species of *Leishmania* seem to utilize amino acids, and perhaps proteins, as energy sources. In sugar-free media, *L. tropica* continues to grow, liberating ammonia. In *L. donovani* it has been reported that a single hexokinase phosphorylates glucose, fructose, mannose, galactose, and glucosamine. This is in sharp contrast to the multiple, highly specific hexokinases of certain flatworms (pp. 78 and 100).

Unlike uninfected tissues, liver and spleen infected with *L. donovani* or *L. tropica* have been reported to carry out aerobic fermentation. Oxygen consumption, as well as aerobic fermentation, was directly proportional to the intensity of infection. Amastigote stages isolated from host tissues showed less intense metabolism than promastigotes from cultures. However, there is reason to suspect that metabolism of the intracellular amastigote stages may be integrated with host cell metabolism, and that actual metabolism of these stages in the host might be quite different from what is measured in salines or in sera.

Under anaerobic conditions the culture forms of *L. donovani* produce lactic and succinic acids as end products of metabolism. Under aerobic conditions the products include lactic, pyruvic, acetic, and succinic acids, as well as carbon dioxide.

Antimonial drugs used in the treatment of leishmaniasis have been shown to inhibit oxygen consumption and the utilization of glucose by *L. tropica* or *L. donovani*. Interestingly the trivalent antimonials were much more powerful inhibitors than pentavalent compounds. It is known that pentavalent antimony is converted to the trivalent form in the host. The exact point of inhibition is not known, but the observations suggest an interference with pyruvate metabolism.

There appear to be real differences in the electron transport mechanisms of the amastigote and promastigote stages. Iron porphyrin compounds are demonstrable at low levels in the intracellular amastigote forms. When placed in culture in vitro, iron prophyrins become readily demonstrable at higher concentrations in about five hours. When examined spectrophotometrically, fully formed promastigotes show absorption bands corresponding to cytochrome *b* and *c* components. Along with this the promastigotes are more sensitive to cyanide than are the amastigotes. The respiration of *L. tarentolae* promastigotes is also inhibited by sodium amytal and antimycin A.

The culture forms of several species of *Leishmania* metabolize various intermediates of the Krebs cycle, and the respiration of *L. tarentolae* is inhibited by malonate. Although the proof is incomplete, the promastigotes appear to have a functional Krebs cycle. Unlike some of the other hemoflagellates, *Leishmania* can apparently oxidize acetate.

Lipid Metabolism

Although the synthesis of long-chain fatty acids commonly proceeds from acetyl-CoA as a starting point, there is some evidence that this may not always be the case in *Leishmania*. When *L. enrietti* was grown in the presence of ^{14}C-labeled stearic acid, significant amounts of radioactivity appeared in stearaldehyde, and in stearic, oleic, linoleic, and linolenic acids; but no label was found in any acid of less than 18 carbons. This was interpreted to mean that unsaturated 18-carbon acids and stearaldehyde were synthesized without degradation of stearic acid and utilization of two carbon compounds. Palmitic acid is also readily desaturated by *Leishmania*. Korn and his colleagues reported that *L. enrietti* and *L. tarentolae* contain α-linolenic acid, which is characteristically found in photosynthetic organisms. In contrast, Meyer and Holz found only γ-linolenic acid in *L. tarentolae*. The γ-isomer is a usual constituent of animal cells. As in the case of many other animals, the composition of the culture medium is known to affect the lipid composition of hemoflagellates. This may be involved in the discrepancies above. In chemically defined media, *L. tarentolae* does not seem to have a lipid requirement, with the exception of a choline requirement when pyridoxal or pyridoxamine are not available (see below). The very high folic acid requirements of hemoflagellates, *Leishmania* included, are probably related to a role of unconjugated pteridines in fatty-acid synthesis.

In a host, *Leishmania* may utilize lipids from the host cell. In tissue culture, chicken macrophages infected with *L. donovani* lose almost all stainable lipid, but it is not known whether this occurs in infected cells in a host.

Nutritional Requirements

Leishmania tropica and *L. donovani* have been reported to require ascorbic acid, whereas the lizard parasites, *L. agamae* and *L. ceramodactyli,* may not require this vitamin. The only species of *Leishmania* that has been cultivated in a chemically defined medium is *L. tarentolae* from lizards. This form required folic acid, an unconjugated pteridine, biotin, pantothenic acid, nicotinamide, riboflavin, thiamine, and either pyridoxine plus choline, pyridoxal, or pyridoxamine.   Ascorbic acid was not required.

In his chemically defined medium, Trager found that *L. tarentolae* required 15 amino acids.   However, the species of *Leishmania* may differ in this regard.   Differences in the ability to synthesize amino acids may be indicated by the finding that *L. donovani* and *L. enrietti* seem to differ sharply in the ability to carry out transamination reactions.   *L. donovani* can use 13 amino acids as amino donors in the pyruvate → alanine system, whereas *L. enrietti* uses only 3 of 13 tested in the same system.   As is true with a number of other hemoflagellates, *Leishmania* seems to require heme compounds.

Speciation of *Leishmania*

The genus *Leishmania* furnishes an example of the difficulties in understanding the evolution of animal parasites.   Specific identification of these parasites has been a thorny problem, as will become apparent.   Not the least of the difficulties is the need for reliable criteria for identification.   Slight differences in morphology in the definitive host may not indicate the profound differences in the biology of the organisms involved.   The students of the leishmanias have been so often concerned with the forms that live in man that they have neglected the forms living in reptiles, which, as Hoare pointed out some years ago, may reflect the probable course of evolution from insect leptomonads to mammalian leishmanias.   Hoare's prediction has basically been borne out, since it has been shown that there are significant antigenic similarities between the forms in reptiles and those in man, and transient infections of mammals have been produced by injecting mammals with forms from reptiles.

One of the difficulties in examining the evolution of the leishmanias rests on the fact that there are so few morphological differences between the so-called species.   For this reason a considerable amount of the data relating to recognition of different forms has been based on the host-parasite relationship and on immunological differences between different forms.   The known species occur only in lizards and in mammals.   Originally the forms found in man were differentiated on the basis of the type of lesion produced by a given parasite.   *Leishmania tropica* produces a skin disease known as Oriental sore; *Leishmania donovani* produces a visceral disease known as kala azar; and *L. brasiliensis* produces South American nonvisceral leishmaniasis.   Russian workers distinguished *minor* and *major* strains of what was referred to as *L. tropica,* on the basis of urban and rural infections, respectively.   These classifications were based on the clinical picture of leishmaniasis in man and variously represented species or subspecies concepts.

It has become clear that the clinical pattern is useful but not infallible in determining the character of the organisms involved in the genesis of these various forms of leishmaniasis. In East Africa the disease during its early stages is attributed to *L. tropica,* but, as it progresses, it becomes a visceral disease that is identical with kala azar in some individuals. The differences in pathology have consequences in transmission, but there are also other differences. East African leishmaniasis is less responsive to treatment with antimonial drugs than is Indian kala azar. Such differences must have a molecular basis.

There has been an enormous confusion in the nomenclature for the species of *Leishmania* associated with human cutaneous and mucocutaneous leishmaniasis of Central and South America. This is by no means clarified at the moment, but some light has been shed on the problem and we may anticipate the unraveling of the difficulties. For a long time the cutaneous form of leishmaniasis in the Americas was attributed to *L. tropica* and the mucocutaneous form to *L. brasiliensis.* Many workers considered these to be two distinct and separate forms of the disease. Other subgroups were suggested from time to time, based mainly on local differences in the disease. Thus *L. peruviana* was named as the agent involved in the disease known as uta in Peru, *L. tropica mexicana* for the cutaneous leishmaniasis of Mexico, *L. tropica guyanensis* for the *Leishmania* in French Guiana, and so on. There is no doubt that leishmaniasis takes different forms in different places. The severe mucocutaneous leishmaniasis of Brazil is quite different from the relatively mild chiclero ulcer of chicle pickers in Mexico. It seems probable that the classical Oriental sore organism, *L. tropica,* was introduced into South America, and there may be reason to believe that the kala azar organism, *L. donovani,* has been introduced. Both of these Old World organisms occur in the Mediterranean and African regions.

Many workers have relied heavily on the results of injecting laboratory animals with a given *Leishmania* for purposes of identification. This approach has proved to be of limited value. Stauber has shown that the responses of various laboratory animals after injection with a single strain of *L. donovani* are highly varied (Fig. 2–14). It now seems likely that newer methods of immunology should be useful in distinguishing the species of *Leishmania.* Earlier attempts that relied on complement fixation or agglutination reactions now also appear to be too crude for this purpose.

The Dyskinetoplastic State

In discussing the trypanosomes and the leishmanias it was pointed out that the kinetoplast contains DNA that differs from the DNA of the cell nucleus. Further, the kinetoplast seems to be the origin for the prolifera-

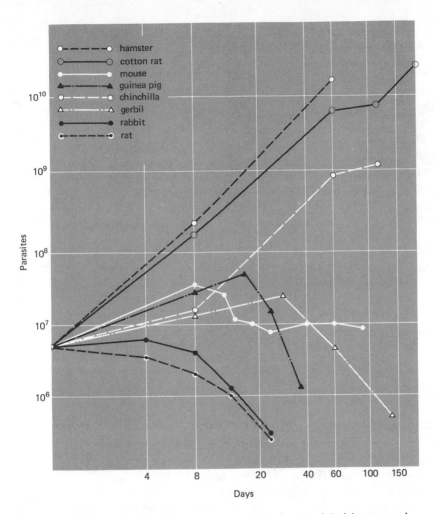

Fig. 2—14. The course of *Leishmania donovani* infections in eight laboratory animals. (After Stauber, 1958.)

tion of the mitochondrion, accompanying the shift from the morphological stages parasitizing the vertebrate to those stages found in invertebrate hosts or in culture outside a host.

When cells are treated with acriflavin, they become what is termed dyskinetoplastic. In this state the kinetoplast DNA virtually disappears, although the remaining part of the kinetoplast structure is retained. In some, but not all, cases, the cell loses the capacity to undergo morphological transformation, mitochondrial proliferation does not occur, NAD diaphorase does not appear, and the cell is not capable of development in the invertebrate host. The loss of DNA renders the kinetoplast unstainable by the usual stains, such as Giemsa's, and in the older literature such cells are referred to as akinetoplastic. Naturally occurring dyskinetoplastic forms have been found from time to time and in these instances are forms that do not undergo cyclic development in an invertebrate host. In some cases there is evidence that the dyskinetoplastic forms multiply more rapidly in the vertebrate host than forms with a normal kinetoplast. This would enhance the probabilities of their survival *if* transmission does not depend on cyclic development in an arthropod. Hoare found dyskineto-plastic strains of *T. evansi* in African camels. These are transmitted mechanically on the contaminated mouthparts of flies. A similar situation pertains with the dyskinetoplastic *T. equinum* of horses in South America.

The failure of the development of cytochrome components and perhaps of the Krebs cycle in dyskinetoplastic hemoflagellates reflects failure of the mechanism for mitochondrial proliferation and seems comparable to the work on bakers' yeast performed by Ephrussi and his colleagues. When yeasts were treated with acriflavin, they gave rise to mutants that grew very slowly, forming so-called "petite" colonies. These mutants had lost the ability to form mitochondria. Careful investigation showed that no mappable mutation had occurred. The failure of such yeast to form cytochrome components and respiratory enzyme elements normally found in mitochondria resulted in an energy metabolism that was fermentative in character under aerobic conditions.

The most important conclusion that can be drawn from the observations on yeasts and hemoflagellates is that uninjured mitochondrial DNA is apparently necessary for the formation of new mitochondria. This means that the mitochondria indeed have some autonomy from the nuclear genetic apparatus of the cell.

In yeast grown under anaerobic conditions, the cells lack fully developed mitochondria. Cytochromes *a* and *b* are not present. When such cells are exposed to oxygen, mitochondria develop and the enzymes for aerobic energy metabolism appear. However, as already indicated, oxygen does not appear to serve such a signal function in the hemoflagellates.

Evolution of Hemoflagellates

In order to study the evolution of this group of protozoans it is necessary to consider briefly some forms that we have not mentioned in preceding pages. Genera related to *Leishmania* and to *Trypanosoma* are symbiotes of insects or of insects and plants. During the past few years new progress has been made in defining the morphology of these various genera. This has proved to be important, since morphological stages and host associations are significant criteria in constructing a picture of evolution. At the same time important studies have been made on the nutrition and metabolism of many of these forms.

Two very different views of evolution in the hemoflagellates were posed many years ago. One school postulated that the trypanosomes arose as parasites of vertebrates and that hematophagous insects subsequently became involved in the life cycles. On the other hand, it was postulated that the hemoflagellates were originally parasites of insects. As some insects developed the bloodsucking habit, some of their flagellates became adapted to life in vertebrate hosts. In the author's view the last hypothesis has most to recommend it. A considerable number of hemoflagellates occur in nonbloodsucking arthropods. In addition, the transmission mechanisms of hemoflagellates show increasing degrees of specialization among the forms found in vertebrates. This is consistent with the concept of an origin in the invertebrates. Among the hemoflagellates living only in insects, the cystic forms escape from the host in the feces and depend on chance to find a new host. Further specialization is found in the forms that parasitize insects and vertebrates, and these are transmitted to the vertebrate when the insect is eaten. There is additional specialization in the forms that are directly transmitted to the vertebrate through injection by a bloodsucking arthropod. Lastly, there are forms that have dispensed with the insect host and are transmitted by contamination of the mucous membranes of the vertebrate.

Unlike the malaria parasites, the hemoflagellate parasites of vertebrates are not limited to a single group of insects. They live in association with flies, fleas, and hemipteran bugs; one species, *Trypanosoma equiperdum,* has completely escaped from an association with an arthropod and is transmitted venereally among horses. When we examine other relationships, the picture seems even more complicated. At least one trypanosome lives in alternate associations with leeches and fishes or amphibians, and some species have been described from marine molluscs. There is clearly a great need for study of the hemoflagellates in lower vertebrates, since much of our thinking has been based on hemoflagellate-mammal associations.

THE TRICHOMONADS

The protozoans of the genus *Trichomonas,* and some closely related forms, are spindle-shaped flagellates, having free anterior flagella, an undulating membrane, and a rod-like axostyle. They move jerkily without the grace of trypanosomes. Many of them live in the gut of vertebrate or invertebrate hosts. Those in vertebrates show somewhat more variation in their location in a host. A given species of *Trichomonas* will tend to live in one, or in only a few, host species. Some of the better known ones are *Trichomonas vaginalis* from the human genitourinary system, *T. tenax* from the human mouth, *T. foetus* from the genitourinary system of cattle, *T. gallinae* from the esophagus and crop of birds, particularly pigeons and doves, and *T. muris* from the cecum of rodents. Some of these forms are of medical and economic significance and have been studied more than other members of the group.

The Life Pattern of Trichomonads

None of the trichomonads are known to undergo any cyclic development, or do they live in intermediate hosts. Sexual reproduction is not known in the group. Transmission from host to host is primarily contaminative in character, although *T. vaginalis* is sometimes, and *T. foetus* usually, transmitted venerally. There is considerable variation in the length of survival outside a host, *T. vaginalis* being most sensitive to extra-host environments and *T. gallinae* being able to withstand freezing in tap water for several days.

Cell Division in Trichomonads

For purposes of discussion the diagrammatic representation of a trichomonad cell in Fig. 2–15 will serve to illustrate the character of the organelles mentioned. The flagellar apparatus in this and in other flagellates is referred to as a *mastigont*. In *Trichomonas* this includes the flagella and undulating membrane, the costa, and the parabasal body and fibril.

The first sign of a cell division is the appearance of a new costa that grows out of the mastigont. The trailing flagellum with its undulating membrane remains associated with the original costa. In this example the three free flagella separate, with two staying together and one going with a second set of organelles. A new parabasal body simultaneously appears. At about this time the nucleus is in prophase. Two sets of mastigonts become more apparent, and a new undulating membrane begins to appear. The axostyle begins to disappear, starting at the anterior end. The nucleus now enters metaphase and, at metaphase, new axostyles begin to appear. As the nucleus attains telophase, the development of the new mastigont is essentially complete, except for the parabasal body that continues to develop. The daughter nuclei separate and lie at opposite sides of the cell, each in association with a mastigont. As the independent locomotor activity of the mastigonts begins, the cell constricts, finally separating into two active cells.

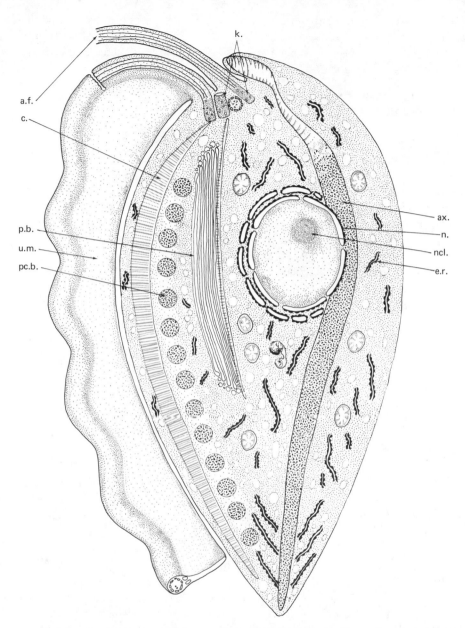

Fig. 2—15. A reconstruction of a trichomonad cell. Abbreviations: a.f., anterior flagellum; ax., axostyle; c., costa; e.r., endoplasmic reticulum; k., kinetosomes; n., nuclear envelope; ncl., nucleolus; p.b., parabasal body; pc.b., paracostal bodies; u.m., undulating membrane.

Nuclear and mastigont division are not always coordinated, and in many forms this results in a variation in the number of flagella. It has been noted that the nature of the culture medium will affect this coordination.

Energy Metabolism

Generally the trichomonads are dependent on carbohydrate as an energy source for growth. They show considerable variability in the specificity of the carbohydrates that can be utilized. *T. vaginalis,* for example, seems to be able to utilize glucose and its polymers for growth but not other sugars. *T. foetus,* on the other hand, grows on a variety of carbohydrate substrates.

Unlike the trypanosomes, the trichomonads store significant quantities of glycogen in the cell. However, all studies of these polysaccharides have been made on materials isolated after treatment with strong alkali, a treatment that has been shown to have a marked effect on molecular weights of animal glycogens. When incubated without an energy source in the medium, several trichomonads have been shown to degrade stored glycogen, and they can live for some days without sugar. Cell extracts of *T. foetus* degrade glycogen to a mixture of hexose phosphates in the presence of inorganic phosphate. The same organism was also shown to carry out uridine diphosphoglucose synthesis in the reaction

Uridine triphosphate + glucose-1-phosphate →
uridine diphosphoglucose + inorganic phosphate.

In addition, *T. gallinae* has an amylomaltase that catalyzes polysaccharide synthesis in the reaction

$$n \text{ Maltose} \rightarrow (\text{glucose})_n + n \text{ glucose}.$$

Most of the trichomonads seem to be anaerobic organisms, although some may be microaerophilic. *T. vaginalis* is very sensitive to oxygen and in aerobic cultures shows a curious failure of normal cell division. Cytoplasmic division does not occur, and multinucleate "monsters" appear. All trichomonads studied have been shown to take up oxygen under aerobic conditions, but this is of dubious physiological significance, since many of the same species grow most luxuriantly in anaerobic cultures. Aerobic respiration is insensitive to cyanide, so it apparently does not involve heavy metal systems.

Several trichomonads have been shown to produce molecular hydrogen, carbon dioxide, and acids as end products of carbohydrate metabolism. In *T. foetus,* most of the acid is succinic and acetic, along with small quantities of lactic and pyruvic acids. There is no net production of carbon dioxide in this case; carbon dioxide fixation appears to be very important, and as in the case of some of the worms (p. 90), the presence of carbon dioxide stimulates carbohydrate metabolism. Some workers have found *T. vaginalis* to produce only lactic acid and very little hydrogen, whereas others have found succinic and malic acids as well, along with larger quantities of hydrogen. Some reasons for differences in such experiments are that *T. vaginalis* shows variations in the pattern of metabolism as a function of the type of culture medium, the age of the culture, and the strain of the organism. It is difficult to identify the probable pattern of metabolism occurring in a host. In spite of this, the author and his colleagues collected gas from a "foaming" case of *Trichomonas* vaginitis and identified hydrogen as a component.

A number of enzymes of the Embden-Meyerhof glycolytic sequence have been identified in trichomonads. Nevertheless, after observing the fixation of radioactive carbon dioxide in the carboxyl carbon of lactic acid by *T. vaginalis,* Wellerson and his colleagues have postulated that a pentose phosphate pathway is important in carbohydrate dissimilation. This requires further investigation.

Malic acid is metabolized by *T. vaginalis* and may be involved as a product of carbon dioxide fixation leading to succinic acid formation. Other intermediates of the Krebs cycle do not seem to be metabolized by *T. vaginalis* or *T. foetus.* On the other hand, *Trichomonas gallinae* was shown to oxidize six intermediate compounds identified with the Krebs cycle. There thus appears to be some variation within the group with regard to the presence or absence of a Krebs cycle. But a word of caution must be inserted. As with the leishmanias, the proof of a *functional* Krebs cycle has not been obtained for any trichomonad.

Anaerobic gas production of *T. vaginalis* is inhibited by cyanide and by carbon monoxide, suggesting that a heavy metal system might be involved in electron transport. However, the aldolase of this organism is a heavy metal system. Thus the effects of cyanide on glucose metabolism are somewhat ambiguous at present. The organisms do not seem to have cytochrome components, but they may have a ferridoxin system. *T. foetus* has a formic dehydrogenase that might be involved in hydrogen production, but attempts to characterize the hydrogen-forming system have failed.

Virulence of Trichomonads

In the past few years there have been several reports that some strains of *T. gallinae* and *T. vaginalis* invade host cells. This invasive capacity seems to be correlated with the relative virulence of different strains.

Earlier reports that trichomonads produce toxins have not been confirmed by later research workers. It was demonstrated that the pathogenicity of a mild strain of *T. gallinae* could be increased by exposing the cells to a cell-free homogenate of highly virulent cells. The effects were abolished by treating the homogenate with deoxyribonuclease, and the change produced by the untreated homogenate was stable for many generations. This was interpreted as a genetic transformation that was DNA-dependent. There was the possibility of course that a viral agent might be involved, but this has been ruled out. Interestingly, it has now been shown that both RNA and DNA are required for this transformation.

Pathogenicity of *T. gallinae* is also modified by cultivation outside the host. One strain that killed pigeons in eight or nine days was no longer lethal after about 40 serial transfers in culture. On re-passage in birds, the organism regained its lethal capacity. When maintained for longer periods in culture, this strain lost its ability to establish itself as a parasite of birds. There are also immunological changes in *T. gallinae* after prolonged cultivation outside a host. There is some evidence that a number of these effects are not simply due to the replacement of virulent forms by avirulent mutants, but may be due to dilution of a cytoplasmic factor that is not replaced in culture media and is of host origin.

ENTAMOEBA HISTOLYTICA: THE DYSENTERY AMOEBA

The amoebae of the genus *Entamoeba* are sarcodine protozoans that move by the formation of pseudopodia. They vary considerably in their biology: *Entamoeba histolytica* is parasitic in primates; *E. invadens* is parasitic in reptiles; *E. coli* is a harmless commensal in man, monkeys, and dogs; and *E. moshkovskii* is found in sewage. Those species living in mammals typically inhabit the large intestine. A notable exception is *E. gingivalis,* which lives in the mouth. *E. histolytica* causes serious disease in man but was not discovered until 1875 and was not named until 1903. It is cosmopolitan in its distribution but occurs with highest frequency in the tropics. Until rather recently there was an astonishing confusion concerning the amoebae included under the name *E. histolytica.* It was long recognized that there were "small and large races" of *E. histolytica.* The small race did not appear to produce pathology in the host. It has since become obvious that many of the forms referred to as "small race" represent a different species, *Entamoeba hartmanni.* However, because this form was commonly referred to as *E. histolytica* in many parts of the world, particularly the English-speaking world, it is quite impossible to construct a meaningful picture of the geographical distribution of amoebic disease in man. We shall say more about these difficulties later.

The Life Pattern of *E. histolytica*

The vertebrate becomes infected with *E. histolytica* when the encysted stage is ingested with food or water. A tetranucleate amoeba emerges from the cyst in the host intestine. Cytoplasmic division ensues, forming typical uninucleate amoebae (*trophozoites*). Cell division may continue in the lumen of the large intestine, and some of the amoebae may invade the

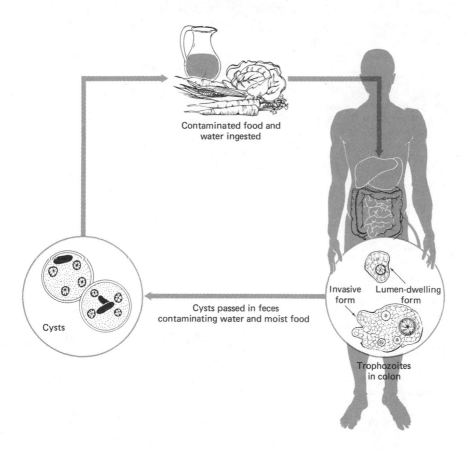

Contaminated food and water ingested

Cysts

Cysts passed in feces contaminating water and moist food

Invasive form

Lumen-dwelling form

Trophozoites in colon

Fig. 2–16. The life pattern of *Entamoeba histolytica*.

intestinal mucosa, producing amoebic ulcers. Only those forms living in the intestinal lumen are capable of rounding up and producing the cysts that pass out of the host in the feces. The life pattern is diagrammed in Fig. 2–16.

Feeding and Digestion

Presumably amobae may take up solutes from the ambient medium, but no data on mechanisms are available. It is well known that solid material is taken in by phagocytosis. In *Entamoeba histolytica,* particulate food, such as red blood cells, bacteria, or starch grains, seems to be taken in with essentially no accompanying fluid. Inside the amoeba, small cytoplasmic vesicles congregate about the ingested particle, and an enlarged food vacuole is formed. It has been postulated that these vacuole-forming vesicles may be lysosomes.

E. histolytica may have digestive enzymes in the cell membrane or secrete such enzymes into the surrounding medium. When amoebae are brought into contact with agar films containing DNA or RNA, hydrolysis of the nucleic acids occurs. Casein, gelatin, or hemoglobin, but not collagen, are also hydrolyzed on such plate preparations. A trypsinlike proteinase, as well as carboxypeptidase, aminopeptidase, and dipeptidase, has been described in homogenates of *E. histolytica.* Both *E. histolytica* and *E. coli* degrade starch when intact cells are placed on starch films, and an amylase has been demonstrated in cell homogenates of *E. histolytica* and *E. invadens.* However, it is also well known that these forms take up starch particles by phagocytosis.

Energy Metabolism of *Entamoeba*

Most work has been concerned with forms that were called *Entamoeba histolytica.* However, the problems of species recognition in this genus are difficult, and some apparently conflicting results obtained in studies of energy metabolism may be due to the fact that different species were studied. In addition, some studies have been carried out with amoebae grown in the presence of bacterial associates. Investigators have attempted to remove associated bacteria by washing amoebae, but the results obtained must be accepted with some reservation. Since the cultivation of *E. histolytica* without bacteria was accomplished by Diamond in 1961, it may be anticipated that future studies will clarify some of the discrepancies.

In 1954, Entner and Anderson found that *E. histolytica* produced lactic acid from the fermentation of endogenous substrates, and in the presence of carbon dioxide some succinate was produced. In 1956, Kun et al. reported that the utilization of glucose was enhanced by the sulfur amino acid cysteine and that carbon dioxide and hydrogen sulfide were produced; also that negligible amounts of lactate were formed and the carbon dioxide was thought to arise from the decarboxylation of pyruvate to acetaldehyde; and further, that alanine was formed as a product of cysteine metabolism. Bragg and Reeves, in 1962, found that the metabolism of labeled glucose resulted in the formation of carbon dioxide, hydrogen, ethanol, acetic acid, traces of lactic acid, and an additional unidentified acid; no hydrogen sulfide was found.

The major storage product of *Entamoeba* is glycogen, and the parasites utilize either starch or glucose from the ambient medium for glycogen synthesis. Entner and his colleagues have shown that different "strains" of *E. histolytica* can be differentiated by their capacity to utilize various sugars.

Several enzymes of the glycolytic sequence have been identified, as well as the enzymes of the Entner-Doudoroff pathway (glucose-6-phosphate → phosphogluconate → triose phosphate + pyruvate). The plethora of different products reported suggest that there may be considerable variation in the terminal steps of carbohydrate degradation in the species or strains of *Entamoeba*. More research is required to sort out the situation.

Who Is *Entamoeba histolytica*?

There seems to be no doubt that thousands and thousands of humans have been subjected to unpleasant and sometimes dangerous treatment for an amoebic disease that was nonexistent. Some physicians, particularly in the United States, adopted the view that anyone with an amoebic infection should be treated. This is very much like treating all patients who harbor *Escherichia coli* in the intestine, just because *E. coli* sometimes produces pathology outside the digestive tract. *Entamoeba histolytica* has been blamed as a causal agent for rheumatoid arthritis, various diseases of the eye, obscure pulmonary diseases, and allergies of various kinds, including migraine, skin rashes and hives, pruritis, and so forth. In some instances these symptoms were allayed by treatment with the antiamoebic drug emetine. However, anyone who has taken a course of emetine treatment might feel better if only in self defense! As Elsdon-Dew remarked, "It is painfully apparent that many 'diagnostically destitute' physicians have seized upon the amoeba as a convenient scapegoat for their inability to find the real cause of the patient's symptoms."* There are indeed well-authenticated cases of amoebic invasion of the liver, brain, skin, cervix, and some other organs, but in such unusual cases the amoebae are found at the site of the pathology. On the other hand, there is no sound evidence that *E. histolytica* produces pathology at remote locations through toxins

---

* "The epidemiology of amoebiasis," *Adv. in Parasitol.* 6(1968):17.

or such.   As an example of a scientific muddle it seems desirable to review briefly the way in which we reached this state of affairs.

Much of the early confusion concerning the status of *Entamoeba histolytica* was initiated by Schaudinn around the turn of the century. Schaudinn named *E. histolytica* and described a life pattern for the organism that was totally erroneous.   Among other things the tetranucleate cyst of *E. histolytica* was given another name.

Between 1910 and 1920 some of this confusion was ameliorated. However, two schools of thought arose concerning the relationships of the minute, invasive, and cystic stages of *E. histolytica* to one another.   Some European workers, notably Kuenen and Swellengrebel, considered the minute stage to be a lumen-dwelling form that did not feed on host tissue and gave rise to cysts appearing in the feces.   On the other hand, Dobell in Britain and Craig in the United States held the view that *all* trophic stages of *E. histolytica* were obligatory tissue-feeding parasites.   In the English-speaking world, this latter view prevailed and led to the widespread tendency to treat all patients even suspected of harboring amoebae.   Dobell subsequently changed his views, but his earlier concept remained prevalent, owing to the fact that it had already been promulgated in textbooks.

During this period and in the following years, various workers noted

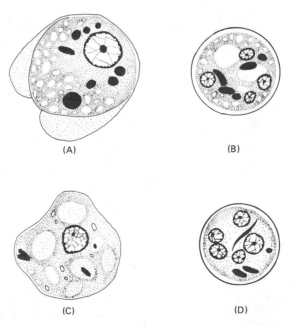

(A)

(B)

(C)

(D)

Fig. 2–17.   Trophozoites (A, C) and cysts (B, D) of *Entamoeba histolytica* and *E. hartmanni.* (A and B) *E. histolytica.* (C and D) *E. hartmanni.* (After Hoare, 1959.)

that those races of *E. histolytica* that produced small (6–8 μ) cysts did not seem to be involved in amoebic disease. Although von Prowazek had named this small amoeba *Entamoeba hartmanni* in 1912, it was widely regarded as a race of *E. histolytica.* Finally in the 1950s, British and American protozoologists began to accept the concept that *E. hartmanni* was a nonpathogenic species rather than a race of *E. histolytica.* Amazingly, some recent American textbooks (e.g., Faust, et al., 1968; Jones, 1967) still fail to make clear the specific distinction between *E. hartmanni* and *E. histolytica,* even though they are demonstrably different in morphology and immunology and probably in geographical distribution. The two species are shown in Fig. 2–17.

REFERENCES

Bishop, A. 1967. Problems in the cultivation of some parasitic Protozoa. *Adv. in Parasitol.* 5:93.

Danforth, W. 1967. Respiratory metabolism. In *Research in Protozoology* 1:201. T. T. Chen (ed.). Elmsford, N.Y.: Pergamon Press, Inc.

Elsdon-Dew, R. 1968. The epidemiology of amoebiasis. *Adv. in Parasitol.* 6:1, 17.

Garnham, P. C. C. 1967. Malaria in mammals excluding man. *Adv. in Parasitol.* 5:139.

Hill, G. C., and W. A. Anderson. 1970. Electron transport systems and mitochondrial DNA in Trypanosomatidae: A review. *Exptl. Parasitol.* 28:356.

Hoare, C. A. 1967. Evolutionary trends in mammalian trypanosomes. *Adv. in Parasitol.* 5:47.

Honigberg, B. M. 1967. Chemistry of parasitism among some Protozoa. In *Chemical Zoology* 1:695. G. Kidder (ed.). New York: Academic Press, Inc.

————. 1968. Some recent advances in our understanding of parasitic Protozoa. *J. Protozool.* 15:223.

Moulder, J. W. 1962. *The Biochemistry of Intracellular Parasitism.* Chicago: University of Chicago Press.

Neal, R. A. 1966. Experimental studies in *Entamoeba* with reference to speciation. *Adv. in Parasitol.* 4:1.

Sadun, E. (ed.). 1969. Experimental malaria. *Military Medicine* (Suppl.).

Vickerman, K. 1971. Morphological and physiological considerations of extracellular blood Protozoa. In *Ecology and Physiology of Parasites.* A. M. Fallis (ed.). Toronto and Buffalo: University of Toronto Press.

As a class of flatworms, the Trematoda show a major common characteristic whereby they live in symbiosis with other organisms. The orders in the class probably have polyphyletic origins from free-living flatworm ancestors. Parasitologists have long been fascinated with the intricate patterns of trematode life histories, and discussing the evolution of some of these complex patterns indeed calls for imaginative interpretation. However, such will not be attempted here. Rather, we will examine a few species simply to illustrate the complexities.

Most of the trematodes living in other animals inhabit the digestive tract or its accessory tubes and cavities, and it seems probable that those species living outside the gut are derived from gut-dwelling forms. The forms to be discussed below have been chosen because we know a great deal about them and can thus compare them with parasites belonging to other animal phyla.

Trematode Parasites

chapter 3

FASCIOLA HEPATICA

This organism, frequently called "the common liver fluke," is a trematode parasite of cattle, sheep, and other herbivorous animals. Like *Ascaris,* the liver fluke has been studied as an exemplary parasite by many thousands of students in elementary zoology courses. The pedagogic choice appears to have been dictated by its large size and availability rather than its value as a typical representative of the trematodes. Its inclusion as an example in this book is based, in part, on the quantity of information available on *Fasciola* and does not constitute an endorsement of the organism as a "typical" form.

The Life Pattern of *Fasciola*

The adult *Fasciola* is found in the bile ducts of various herbivorous mammals and is a flattened, leaf-shaped organism (Fig. 3–1). Adult worms are hermaphroditic, but cross-fertilization between individuals is probably the commonest mode of sexual reproduction. The shelled embryos undergo cleavage before leaving the parental uterus and pass out of the host in the feces.

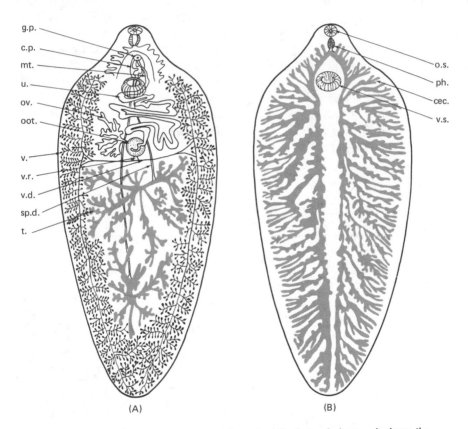

Fig. 3–1. The adult *Fasciola hepatica* from the bile ducts of sheep. A shows the reproductive systems only. B shows the digestive system only. (Redrawn from Chandler and Read, 1960.) Abbreviations: cec., cecum; c.p., cirrus pouch; g.p., gonopore; mt., metraterm; oot., ootype; o.s., oral sucker; ov., ovary; ph., pharynx; sp.d., sperm duct; t., testes; u., uterus; v., vitellaria; v.d., vitelline duct; v.r., vitelline reservoir; v.s., ventral sucker.

Under favorable environmental conditions (water and temperature), the embryo develops into a ciliated larval form, the *miracidium* (Fig. 3–2), which hatches and swims about with the aid of its cilia. If the miracidium encounters a suitable snail species within a few hours, it penetrates the snail, shedding its ciliated cells, and transforms into a saclike *sporocyst* (Fig. 3–3). As the sporocyst grows, new larval forms develop within it. These forms, called *rediae* (Fig. 3–3), leave the sporocyst and typically move to the digestive gland (hepatopancreas) of the snail host. The rediae show more evidences of differentiation than are obvious in the sporocyst. As the primary rediae grow, new daughter rediae develop within them. Eventually rediae give rise to another embryonic type, the *cercaria* (Fig. 3–4). This form has a tail and, after leaving the rediae, eventually escapes from the snail into the surrounding water. In a matter of minutes, or at most a few hours, the cercaria encysts on a blade of grass or other vegetation and becomes a *metacercaria*.

This encysted form is capable of infecting a mammalian host if and when it is inadvertently ingested along with a meal of vegetation. The pattern of the life history shows a series of adaptations for ensuring the perpetuation of *Fasciola* as a parasite of grazing animals feeding in boggy pasture lands (Fig. 3–5).

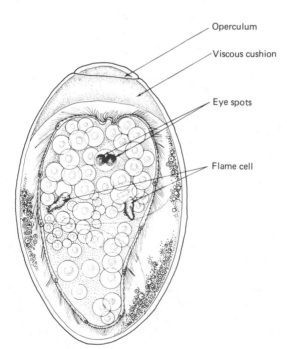

Operculum

Viscous cushion

Eye spots

Flame cell

Fig. 3–2. The fully developed miracidium of *Fasciola*, just before hatching from the enclosing shell.

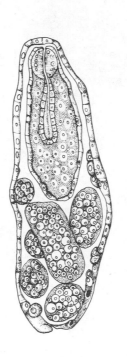

Fig. 3—3. The sporocyst (left) and redia (right) of *Fasciola hepatica*. (Redrawn from Thomas, 1883.)

The "Egg"

The formation of the shelled trematode embryo, incorrectly but commonly referred to as an egg, is a complex process, and some understanding of the reproductive apparatus is required (Fig. 3–6). The male and female systems open to the outside through a common genital pore. The male portion of the system involves paired, branched testes connected through a common sperm duct to a seminal vesicle and thence to a cirrus. The female system consists of a single highly branched ovary, an oviduct, the paired, branched vitelline glands, a chamber in which shell formation is initiated (ootype), and some glands usually referred to as Mehlis' glands. Joined to this complex is the uterus that connects with the genital pore by way of a metraterm.

The biflagellate sperm of *Fasciola* is thought to fertilize the ovum in

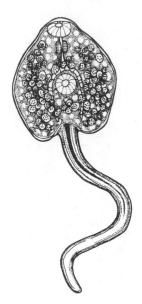

Fig. 3—4. The free-swimming cercaria of *Fasciola hepatica*. (Redrawn from Thomas, 1883.)

the ootype. At about the same time, 30 or so vitelline cells are released from the vitelline reservoir (see Fig. 3–1), and the ovum plus the accompanying vitelline cells receive the Mehlis' gland secretion, a lipoprotein. The vitelline cells then secrete granules of material that appear to be the precursor material for the shell. This material encloses the ovum and the accompanying vitelline cells; cleavage also begins.

The newly formed soft shell and the vitelline cells contain proteins, phenols, and the enzyme phenolase (phenol oxidase). As the shelled embryo passes along the uterus, the shell hardens; and as it does so, it loses its reactivity in tests for free phenols, basic proteins, and phenolase. The evidence suggests that the tanned protein of the shell is formed by what is termed quinone tanning. In such tanning, a quinone, formed by oxidation of a phenol, reacts with adjacent protein chains to form a highly stable cross-linked structure. This may be described as follows:

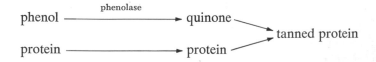

Interestingly, serotonin (5-hydroxytryptamine) inhibits the phenolase of *Fasciola;* this same amine seems to serve a quite different function in the carbohydrate metabolism of the fluke (see p. 67).

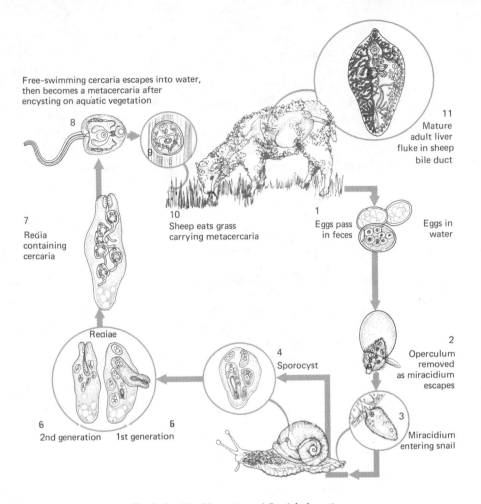

Free-swimming cercaria escapes into water, then becomes a metacercaria after encysting on aquatic vegetation

8

9

11
Mature adult liver fluke in sheep bile duct

7
Redia containing cercaria

10
Sheep eats grass carrying metacercaria

1
Eggs pass in feces

Eggs in water

Rediae

4
Sporocyst

2
Operculum removed as miracidium escapes

6
2nd generation

5
1st generation

3
Miracidium entering snail

Fig. 3—5. The life pattern of *Fasciola hepatica*.

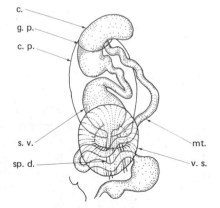

Fig. 3—6. The terminal reproductive organs of *Fasciola*. Compare with Fig. 3—1 for orientation. Abbreviations: c., cirrus; c. p., cirrus pouch; g. p., gonopore; mt., metraterm; s. v., seminal vesicle; sp. d., sperm duct; v. s., ventral sucker.

c.

g. p.

c. p.

s. v.

sp. d.

mt.

v. s.

Development of the Miracidium

The shelled embryos of *Fasciola* leave the vertebrate host in the feces, and although development remains arrested while the embryos remain in the feces, they may survive in wet fecal material for several months. If washed free, the development of the embryo proceeds. Between 10 and 30° C, the rate of embryonic development is a function of temperature, being completed in eight days at 30° C. The availability of oxygen is also a factor; embryos kept under aerobic conditions develop in about one-fifth of the time required by embryos deprived of oxygen.

The fully developed miracidium is a more or less conical little animal covered with a ciliated epithelium. It does not look at all like an adult trematode but does show resemblances to some free-living flatworms (Fig. 3–2).

The miracidium has well-developed eye spots, which may be primary receptors involved in the process of hatching. If kept in the dark, few miracidia hatch, but illumination rapidly induces hatching. There is some evidence that light in the blue and violet portion of the spectrum triggers the release of a proteolytic enzyme that attacks a collar of untanned protein holding the operculum of the shell in place. This releases the operculum except at one point. A "viscous cushion" in the opercular end of the shell cavity appears to swell just before hatching occurs, but the chemical basis of this swelling phenomenon is not understood. As the operculum is loosened, the miracidium becomes quite active and escapes from its enclosing shell.

During the development of the shelled embryo of *Fasciola,* the glycogen content drops from about 32 percent to 15 percent of the dry weight at the time of hatching. During this same period, oxygen consumption increases by about fivefold. The apparently beneficial effects of oxygen on development suggest that the embryonic stages differ from the adults in the significance of aerobic metabolic pathways. However, no detailed studies with any trematode have been made on this point.

The Miracidium

The active free-swimming miracidium has a transient existence. If it does not contact an appropriate mollusc within a few hours, it dies. The miracidium is ordinarily considered to be a nonfeeding stage. However, it has been shown that the miracidium will metabolize glucose added to the suspending medium, and it would appear that this stage could thus feed on organic solutes. No studies seem to have been made on the survival of miracidia in the presence and absence of soluble nutrients. There is evidence that the miracidium may not have a functional tricarboxylic acid cycle. When ^{14}C-glucose was added to the suspending medium, labeled intermediate compounds associated with glycolysis were recovered from the tissues of the trematode. No intermediates of the tricarboxylic acid cycle were recovered. When incubated with ^{14}C-succinic acid, the only labeled intermediate recovered was malic acid.

Miracidia, in general, seem to react to light (see p. 72) and may show chemotropic responses. However, there seems to be little specific information on the reactions of *Fasciola* miracidia to physical factors. The fact that in the laboratory, *Fasciola* is able to infect five of six British snails of the genus *Lymnaea* may suggest that the chemical attraction of snail hosts is relatively broad. However, of these, only *Lymnaea truncatula* can be consistently infected as adult animals. The remaining snail species can be infected only during very early life. This may not be related to the chemotropic responses of the miracidium but to the reactions of the snail to the developing worm (discussed below).

The Sporocyst and Redia

The entrance of the miracidium into a snail is marked by the loss of the ciliated outer layer of cells and by cytolytic effects on the host tissues. There is circumstantial evidence that entry is assisted by worm secretions and is not a simple mechanical process. When the ciliated cells are shed, the resulting organism is a sporocyst.

Germ cells in the sporocyst give rise to a different form, the redia (Fig. 3–3). A second redial generation may be produced, and it gives rise to the cercaria, which subsequently leaves the molluscan host. Virtually nothing is known of the physiology of the intramolluscan stages of *Fasciola*. There is evidence in a number of trematode infections that host tissue deposits of polysaccharide are depleted in the vicinity of the parasites. Although there have been speculations on the mechanism of such effects, the basis of this chemical pathology is unknown. It may be suggested that it involves the production of serotonin (5-hydroxytryptamine), a compound that may regulate glycogenolysis through the production of cyclic AMP in mollusks.

Reaction of the Snail Host

Several snails can be infected with *Fasciola hepatica*, but they do not support the full development of the parasite. In some snail species, development of rediae occurs, but cercariae are not produced. In other trematode infections, host cell infiltration and walling-off of developing parasites have been described, and such phenomena probably occur in *Fasciola* infections. There is the possibility that some hosts may not furnish specific chemical signals for development, although there is no direct evidence for this information deficiency. The fact that the snail *Lymnaea glabra* commonly supports the development of not more than 30 rediae, whereas 160 or more develop in *L. truncutula,* may suggest a difference in the capacity to support worm development rather than host reaction.

Young snails infected with *Fasciola* may become stunted and exhibit distortion of the shell, and there is a decrease in the lipid content of the hepatopancreas. There is some evidence that snails already infected with *Fasciola* may become refractory to further infection with this or other trematode species. Nothing seems to be known of the basis for this acquired immunity.

Factors Affecting Development in Snails

Not surprisingly, the rate of development in the snail host is a function of the environmental temperature. Kendall reported a curious effect of temperature on the development of *Fasciola:* In snails maintained at laboratory temperatures, daughter rediae were not produced, the first generation of rediae giving rise to cercariae. On the other hand, in snails exposed daily to a period of low temperature (4–5° C), daughter rediae were produced from rediae.

The state of snail nutrition also affects development of *Fasciola.* The number of parasites developing is a function of the amount of food obtained by the snail host. This is in contrast to the predisposition to develop large numbers of parasites often seen in undernourished mammalian hosts. This effect of food availability on development in the mollusc hosts is probably the explanation for the observation that fewer trematodes develop in small snails than in large ones, although snails of all sizes may become infected.

The Cercaria and Metacercaria

The cercariae of *Fasciola hepatica* leave the molluscan host in what has been described as a "passive" process. They congregate in the perivisceral space around the distal portion of the gut. An area of the snail's tegument at the anus becomes weakened and protrudes. When the snail closes the pneumostome, cercariae are extruded to the outside through this thin temporary papilla. Some other species of trematodes seem to differ in the mechanism for escaping from the molluscan host.

The cercaria of *Fasciola* probably has a relatively short free life and encysts directly on vegetation. The process of encystment occurs as follows: The cercaria attaches itself to the substratum by the ventral sucker, and the body becomes flattened. It then adheres peripherally and the body contracts, releasing the outer cyst layer. During these early stages of encystment the tail breaks off from the body. The production of cyst material continues and finally consists of four layers, which, from the outside in, are composed of a tanned protein layer, a fibrous layer, a muco-polysaccharide layer, and a laminated keratin layer. The components of this final layer are formed in the cercaria before encystment, the keratin being held in rolls that are unrolled during the late stages of encystment (Fig. 3–7).

The metacercariae, encysted on vegetation, may live for as long as a year at low temperatures, or for as short a time as two or three weeks at 25° C. If they are ingested by a host, the outer cyst wall is reported to

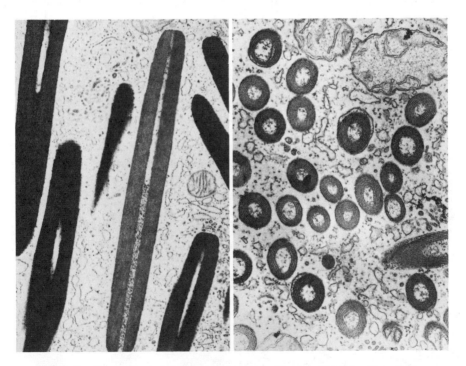

Fig. 3–7. Ultrastructure of the rolls of keratin in the cercaria of *Fasciola*. These are extruded during excystment. (Left) Longitudinal section. (Right) Transverse section. (Courtesy of Dr. Keith Dixon.)

be removed mechanically, presumably by mastication in the usual herbivorous hosts.

Establishment in the Vertebrate

The final excystment in the host requires an elevated temperature (38° C), low redox potential, a high CO_2 tension, and the presence of bile. It is not clear as to whether the stimulating effect of bile is due to the salts of bile acids, as in the excystment of juvenile tapeworms, or to some other constituent. The factors necessary for the excystment of *Fasciola* metacercariae seem to serve as stimuli to activate the worms, but the mechanism is not clearly understood. It may involve secretion by the worm of an enzyme that attacks the mucopolysaccharide components of the cyst.

Since the eventual habitat of adult *Fasciola* is the biliary passages, it might be guessed that the young excysted worms would migrate from the intestine up into the bile ducts. Not so! Amazingly, the young worms penetrate the wall of the gut and pass into the peritoneal cavity. (Some other live flukes, such as *Dicrocoelium,* do migrate from the gut directly into the bile ducts.) In the body cavity the young worms seem to move about rather aimlessly until they encounter the capsule of the liver. On contact with the liver, the flukes burrow into it, devouring liver parenchyma as they move. This period of wandering in the liver is marked by considerable growth of the parasite and, not surprisingly, by considerable reaction on the part of the host. Although a controversy has existed concerning the food of *Fasciola,* it seems clear that at this stage the worm eats the tissues encountered and appears to show little discrimination in food. During the movements of the flukes in the parenchyma of the liver, there are tissue reactions, including hyperplasia of the bile duct epithelium with which the flukes have not yet come in contact. Eventually (three to seven weeks after infection) the bile ducts are invaded and the worms remain there, feeding on bile duct epithelium, components of the bile itself, and probably occasionally on blood liberated by feeding activity. Egg production begins as early as five weeks after ingestion by the mammalian host.

Feeding of Fasciola

Some workers have been impressed with the concept that *Fasciola* is a bloodsucking parasite during its life in the bile ducts. Studies on the amount of radiochromium acquired by flukes in hosts into which Cr[60]-tagged red cells had been injected showed that the flukes obtained the equivalent of 0.18 to 0.2 ml of blood per fluke per day. This amount of blood would contain at most about 2.0 micromoles of glucose. Since there is evidence suggesting that the worms require at least 50 to 100 times this amount of glucose, it seems unlikely that blood is the major source of food for the adult worms in the bile ducts.

In addition to ingesting solid (cellular) and liquid food, *Fasciola* also feeds by absorption through the body surface. Worms with the gut opening ligated take up glucose from the suspending medium at rates similar to those of unligated animals. However, there is good evidence that sugar transport occurs by a mediated process showing chemical stereospecificity. Amino acids are not absorbed by mediated processes through the body surface. Even so, the worm may obtain significant amounts of amino acids from the bile of the host; the free amino acid content of bile from infected hosts is more than ten times higher than that from uninfected animals.

It should be pointed out that the so-called "cuticle" of *Fasciola* is not a nonliving cuticular layer. It is composed of syncytical extensions of underlying cells, and although the tegument lacks nuclei, it is well supplied with mitochondria (Figs. 3–8 and 3–9). The tegument shows phosphatase activity, as is common to many absorptive surfaces, although the role of the phosphatases remains obscure.

The evidence that *Fasciola* utilizes the gut in a digestive capacity is largely circumstantial. Proteolytic enzymes found in extracts of the animal cannot with certainty be identified with digestive function. *Fasciola* has been shown to digest labeled serum albumin, and peptones and coagulated protein have been found in dejecta of the fluke. These suggest a digestive function. Also long, sometimes looped, microvilli found in the gut of *Fasciola* are suggestive of a secretory-absorptive function. Further, extracorporeal digestion of gelatin has been demonstrated with intact specimens of *Fasciola,* although it has not been shown conclusively that the protease involved is derived from the digestive tract of the worm.

Metabolism of Adult *Fasciola*

The tissues of *Fasciola* are rich in glycogen, and carbohydrate seems to be the major substrate for energy metabolism. About 16 percent of the dry weight of the worm is ethanol-precipitable carbohydrate (glycogen), and under overnight starvation conditions, a third to two-thirds of this stored polysaccharide disappears. The polysaccharide is mainly in parenchymal cells, and recent studies with the analytical centrifuge have shown that it is mainly composed of two molecular species having molecular weights of about 50×10^6 and 450×10^6. If the intact worm is furnished with glucose, it consumes it at the rate of about 150 μmoles/g/hr, and a

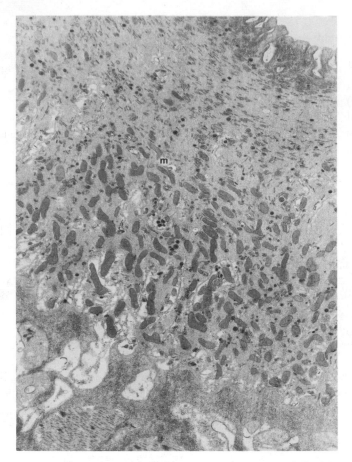

Fig. 3—8. The outer syncytial portion of the tegument of *Fasciola hepatica.* Note the numerous mitochondria (m). (Electron micrograph furnished by Dr. Lawrence Threadgold, The Queen's University, Belfast.)

considerable portion of the sugar metabolized is excreted into the suspending medium as fatty acid metabolites. Propionic and acetic acids, in a ratio of 3:1, account for 80 to 95 percent of the acid produced. A small amount of lactic acid is produced, and some higher fatty acids have been reported as end products of metabolism.

The mechanisms of intermediary metabolism involved in carbohydrate metabolism have been examined in *Fasciola.* Also a number of enzymes of the Embden-Meyerhof sequence have been demonstrated. Interestingly, the worm appears to lack pyruvate kinase, which would catalyze the reaction

Phosphoenolpyruvate + ADP → Pyruvic acid + ATP.

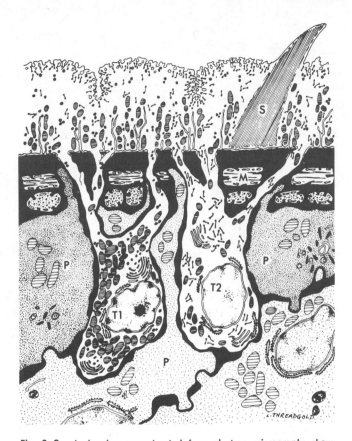

Fig. 3–9. A drawing reconstructed from electron micrographs showing the organization of the outer tegument of *Fasciola hepatica*. The tegument is composed of syncytial extensions of deeper lying tegumentary cells. Abbreviations: S = tegumentary spine; T1 = a Type 1 tegumentary cell; T2 = a Type 2 tegumentary cell; P = parenchymal cell; M = muscle. (Original figure furnished by Dr. Lawrence Threadgold, The Queen's University, Belfast.)

Thus pyruvate and its reduction product lactate do not seem to be formed directly from phosphoenolpyruvate. Rather, the major pathway involves carbon dioxide fixation to form oxaloacetate and subsequent reduction to malate. Pyruvate may then be formed by decarboxylation of malate, or succinate may be formed by reduction of malate with fumarate as an intermediate. The postulated reactions are summarized in Fig. 3–10. From the ratios of products excreted by the intact worm, it would seem that the pathway through succinate (shown in Fig. 3–10) is the favored one.

Serotonin (5-hydroxytryptamine) seems to play a role in the tissues of *Fasciola* comparable to that played by adrenalin in mammalian tissues. Serotonin stimulates carbohydrate metabolism in the intact fluke and has

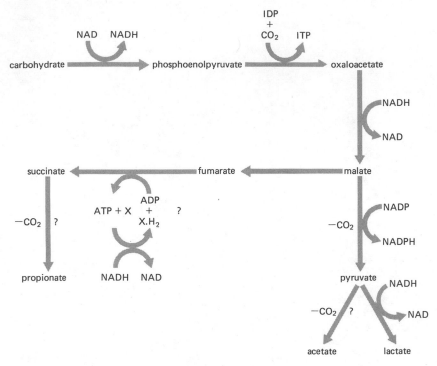

Fig. 3—10. Proposed scheme for degradation of carbohydrate in *Fasciola*. (Taken in part from Prichard and Schofield, 1968, *Biochim. Biophys. Acta* 170:63.)

been shown to have an effect on the synthesis of cyclic 3,5-adenosine monophosphate and to stimulate *Fasciola* phosphofructokinase, which catalyzes the reaction:

Fructose-6-phosphate + ATP ⇌ fructose-1,6-diphosophate + ADP.

Phosphofructokinase appears to catalyze the rate-limiting step in the sequence from glucose to phosphoenolpyruvic acid. The increased quantity of acid produced by serotin-stimulated flukes is mainly lactic acid. As indicated (p. 58), serotonin also affects phenolase in *Fasciola*.

Since the oxygen consumption (80 μmoles/g/hr) would account for the complete oxidation of less than 10 percent of the glucose consumed, it is plain that carbohydrate metabolism is predominantly anaerobic in character. This is further substantiated by the fact that the presence of oxygen seems to have no significant effect on the rate of utilization of endogenous stored carbohydrate.

A further comment concerning oxygen consumption may be made. Although there are no data indicating that aerobic metabolism is of great significance in the economy of *Fasciola*, the worm shows evidences of a generalized type of respiration. Respiration is sensitive to cyanide and carbon monoxide, and the spectra of cytochromes *b* and *c* have been identified in the tissues. *Fasciola* has a very low catalase content. A hemoglobin has also been reported in *Fasciola,* although it has yet to be shown that this is not of host origin. There is circumstantial evidence for the mechanisms for complete oxidation of carbohydrate. Several intermediates of the tricarboxylic acid cycle were recovered from the tissues of flukes incubated with ^{14}C-glucose or ^{14}C-succinic acid. Most of the enzymes of the tricarboxylic acid cycle have been identified in the tissues of *Fasciola hepatica.* NAD-specific isocitrate dehydrogenase has not been found, and the tissues contain very low levels of aconitase and NADP-specific isocitrate dehydrogenase. It seems probable that the tricarboxylic acid cycle functions in anabolic processes, involving the generation of carbon skeletons, rather than having a significant role in energy utilization. Future study may show that the adult *Fasciola* is a microaerophilic animal rather than a true anaerobe.

There is no indication that *Fasciola* can utilize fats as an energy source. Extracts of the worm show very low lipase activity. Also no data are available to decide whether *Fasciola* utilizes amino acids as energy sources. However, the worms have an active nitrogen metabolism. Ammonium ion and urea occur in the tissues and are presumed to be excretory products. There is evidence for a urea cycle, since the worm has arginase and ornithine transcarbamylase, but a urease is lacking. Carbon dioxide fixation into urea has been demonstrated, and there is an increase in urea formation when ornithine is added to preparations of fluke tissue. A number of transaminations can be effected by *Fasciola,* and the occurrence of various amines in the tissues suggests that amino acid decarboxylations can take place, although no examination of the worm for these enzymes seems to have been made.

Hormones in *Fasciola*

Serotonin has already been mentioned as having adrenalinlike activity in *Fasciola,* insofar as effects on metabolism are concerned. It is not known whether serotonin acts as a humoral component of the nervous system, as it is reported to do in some other animals. The tissues of *Fasciola* contain an acetylcholinelike substance and cholinesterase; the specificity of the latter for acetylcholine has not been established, but it has been suggested that these components are related to activity of the nervous system.

Insulin has been reported to have an effect on the carbohydrate metabolism of *Fasciola.* However, studies by Buist and Schofield and those performed in the present author's laboratory have failed to confirm this.

SCHISTOSOMA MANSONI:
THE BLOOD FLUKE

The blood flukes, trematodes of the genus *Schistosoma,* are important parasites of humans. They differ from most trematodes in that they inhabit the blood vascular system and have separate sexes rather than being hermaphroditic. In some areas, such as mainland China, schistosomiasis is of general significance to the public health. The human diseases produced by three species of *Schistosoma* are insidious and debilitating in character. These chronic, drawn-out diseases contribute to death at what Americans would call an early age—30 to 40 years—but this is well past the minimum age for human reproduction. Thus schistosomiasis affects the death rate without seriously affecting the birthrate or reproductive potential of infected human populations. Egypt and other parts of Africa, Southern Asia, and tropical America carry burdens of schistosomiasis, along with other medical and economic problems.

The Life Pattern

Manson's blood fluke lives in the small mesenteric veins of the mammalian host, and the female worm deposits shelled embryos in these small blood vessels. Many of these embryos pass out of the veins into the tissues of the intestinal wall, finally reaching the lumen of the gut. The fully developed shelled embryo leaves the host body in the feces. Upon contact with water of low salt content the *miracidium* emerges from the shell as a free-swimming organism, which may survive for up to 24 hours. For continuation of life, the miracidium must actively penetrate a snail host during this single day of free life. After penetration the ciliated outer epithelium is shed, and the worm becomes a tubular *sporocyst* that grows to about 1 mm in length and begins producing daughter sporocysts within two to three weeks. Daughter sporocysts emerge from the mother sporocyst, undergo growth, and make their way through the molluscan viscera to the digestive gland. Each daughter sporocyst in turn gives rise to numerous fork-tailed *cercariae.* These cercariae begin to leave the snail about six weeks after entry of the miracidium. The cercaria is a free-swimming form that must enter the skin of a mammalian host within two or three days or die. During penetration of mammalian skin the cercarial tail is shed. Within a few hours the young blood fluke, or *schistosomula,* enters the blood system. It tarries for a day or so in the pulmonary circulation, and about six days after entering the mammal it is found in the circulatory system of the liver.

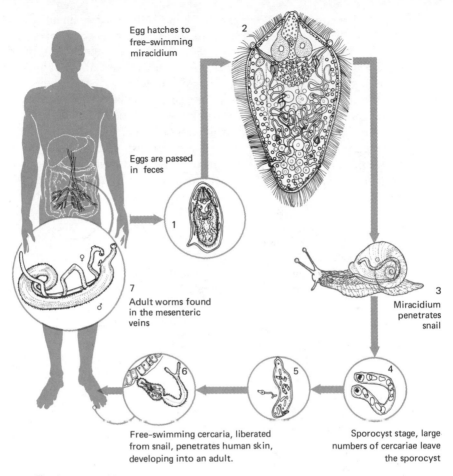

Egg hatches to
free-swimming
miracidium

Eggs are passed
in feces

1

7
Adult worms found
in the mesenteric
veins

3
Miracidium
penetrates
snail

6

5

4

Free-swimming cercaria, liberated
from snail, penetrates human skin,
developing into an adult.

Sporocyst stage, large
numbers of cercariae leave
the sporocyst

Fig. 3—11. The life pattern of *Schistosoma mansoni.* See text for description.

The worm grows considerably here, and about two and one-half weeks later it migrates to the mesenteric veins. Mating and egg production ensue, and the shelled embryos are produced about 40 days after entering the vertebrate host. This pattern is summarized in Fig. 3–11.

Physiology of the Miracidium

The miracidium is fully developed when the shelled embryo of *Schistosoma* leaves the body of the vertebrate host. The miracidium does not "hatch" in saline solutions that are isosmotic with the body fluids of the vertebrate. Further, hatching is inhibited at 37° C. Dilution of the suspending medium to about one-sixth the salinity of mammalian tissue fluids and lowering of temperature to 28° C are necessary for extensive emergence of miracidia to occur. Hatching is also enhanced by light. Unlike *Fasciola,* the shells of *Schistosoma* embryos are not operculate, and the mechanism of hatching does not appear to have been studied. The short life of the swimming miracidium suggests that it utilizes its endogenous reserves, but no data on this point seem to be available.

Miracidia of schistosomes exhibit phototactic, thermotactic, and geotactic responses. *S. mansoni* miracidia show responses to light, but these are overridden by negative geotropism. The relationships of temperature and light intensity on the phototactic responses of *S. japonicum* are shown in Table 3–1. In general, the responses of the schistosome miracidium

TABLE 3–1. The effect of temperature and light intensity on phototaxis of *Schistosoma japonicum* miracidia.

Light intensity				Temperature (° C)						
(LUX)	15	18	20	22	23	24	25	26	28	30	34
4500	+	−	−	−	−	−	−	−	−	−	−
2500	+	±	−	−	−	−	−	−	−	−	−
2000	+	±	±	−	−	−	−	−	−	−	−
1000	+	+	±	±	±	±	−	−	−	−	−
500	+	+	+	+	+	+	±	±	−	−	−
250	+	+	+	+	+	+	+	+	±	−	−
100	+	+	+	+	+	+	+	+	+	−	−
50	+	+	+	+	+	+	+	+	+	±	−
25	+	+	+	+	+	+	+	+	+	+	−
10	+	+	+	+	+	+	+	+	+	+	±

will tend to bring it into proximity with the appropriate snail host species.

A number of investigators have examined the question of the attraction of schistosome miracidia to the molluscan host. There is some evidence that a chemotaxis is involved, but it does not appear to be highly specific, miracidia being attracted to snails that will not serve as satisfactory hosts for subsequent development. Using the flying spot microscope, it was shown that whereas miracidia swim in straight lines in pure water, they show erratic turning behavior in the presence of snail extracts (Fig. 3–12). There is some evidence that the substances responsible for the chemotaxis are fatty acids. From the above, a tentative hypothesis on host finding would include the general responses to light, gravity, and temperature, which would bring the miracidium within range of the host snail, and at this point chemotactic responses would come into play.

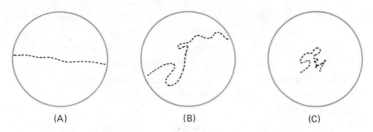

(A)   (B)   (C)

Fig. 3—12. Two-second exposures of the movements of *Schistosoma mansoni* miracidia as revealed by the flying spot microscope: (A) in water; (B) in filtered extract of whole ground host snail, *Taphius glabratus;* (C) whirling behavior frequently exhibited when first encountering host extract. (After Davenport, 1966.)

The penetration of *S. mansoni* into the snail host has been described as a "boring" action, with the attenuated apical papilla assisted by cytolytic secretions from the glandular "gut." The ciliated epithelium is said to be shed after penetration of the snail. The miracidial structures (previously termed "penetration glands") are considered by Wajdi (1964) to produce adhesive and lubricatory substances that are involved in penetration.

Development in a Snail

After penetration there is a period of relative inactivity. The ciliated epithelium, characterizing the miracidium, is shed, and metamorphosis into a mother sporocyst occurs. In a satisfactory host, no host tissue response is observed. On the other hand, in unsuitable snail strains there may be a marked tissue reaction. A rapid cellular infiltration around the newly entered miracidium is followed by walling off by fibroblasts and destruction of the young worm within 48 hours. In a suitable host, little host reaction occurs until five days or more after infection, at which time the growing mother sporocyst may cause occlusion of blood sinuses. During the next few days daughter sporocysts begin developing in each mother sporocyst. These leave the exhausted mother sporocyst and migrate to the digestive gland. The growing daughter sporocysts produce changes in the digestive gland tissues, including cytopathology of the epithelial lining of the digestive gland and a lowering of the glycogen reserves of the host.

Cercariae develop in each daughter sporocyst. The cercariae of *S. mansoni* migrate from the spaces of the digestive gland by way of the blood sinuses to the area of the mantle collar and pseudobranch. Here they penetrate the snail's skin, emerging in the water. A small pair of basophilic glands can be demonstrated in the anterior region of unemerged cercariae; after leaving the snail these are no longer demonstrable, and it has been surmised that they function in the emergence from the snail host. There may be a considerable amount of host cellular activity associated with the movement of cercariae through the tissues of the snail, and in heavily infected molluscs, egg production may be completely suppressed, although there is no degeneration of the gonad. This "physiological castration" may have an endocrine basis.

The number of cercariae resulting from the infection of a snail by a single miracidium may exceed 200,000, although it has been observed that the number produced in infections by several miracidia does not increase proportionally.

Behavior and Physiology of the Cercariae

The active escape of *S. mansoni* cercariae through the intact skin of the snail host is affected by the external environment. The larval worms generally emerge in daylight, and their appearance can be delayed by keeping the host in the dark. Not unexpectedly, cercarial emergence is also inhibited by low temperature.

The free cercariae swim and rest intermittently and are presumed to rely completely on the contained food reserves as their energy source. Although little is known of the metabolism of schistosome cercariae, there is some evidence that aerobic metabolism is essential. If deprived of oxygen, *S. mansoni* cercariae quickly become nonmotile and soon die. This is of some interest when compared with the metabolism of the adult worm (see p. 77). It would be of considerable interest to determine whether lipids may serve as an energy source for schistosome cercariae.

The taxes of *S. mansoni* have been inadequately examined. The larvae react positively to light and to a temperature gradient, but precise data are not available.

Penetration of the Vertebrate Host

Upon initial contact with vertebrate skin, the cercaria moves over it in an exploratory fashion, attaching alternately by the oral and ventral suckers. At each point of oral sucker attachment the young worm deposits droplets of secretion from the postacetabular glands (see Fig. 3–11). This material swells and becomes sticky and serves as a connecting point for each subsequent attachment of the ventral sucker. Eventually the cercaria stops its exploration and becomes anchored in a droplet of the secretion that has become more sticky. The worm now orients itself perpendicularly to the skin surface and exhibits vigorous movements of the body that squeeze more secretion from the postacetabular glands. At the same time the muscular oral sucker is everted and withdrawn repetitively, bringing the cuticular tips of the acetabular gland ducts in contact with the skin and allowing deposit of the acetabular gland secretions into the entry site. Gradually the young worm begins to pierce the horny layer of the skin. The evidence available suggests that the thin and tough horny layer is softened by the alkaline mucus secretion of the postacetabular glands.

As the oral end of the worm breaks through the horny layer of the skin the young schistosome turns and follows the stratum lucidum, or keratogenous zone, of the skin. At this point the worm begins to show contractions of the musculature around the preacetabular glands, and secretions from these glands are poured into the surrounding tissue. At about this time the entire body of the worm is in the skin, and the tail has been shed from the body. The pre- and postacetabular glands are exhausted of their secretions, which are not subsequently renewed. At this point the worm may remain quiescent for several hours.

In addition to the strenuous events of penetration just described, some important physiological alterations apparently occur in the worm. For example, while Ringer's solution and normal serum have toxic effects on free-swimming cercariae, these substances are without obvious effect on forms from the skin, now called schistosomulae. On the other hand, cercariae live happily in water, whereas schistosomulae rapidly die in water; cercariae form a curious membranous envelope in immune serum, whereas schistosomulae do not. These changes in responses of the cercaria to those of the schistosomula may occur in a 15-minute period and obviously represent some fundamental alterations of worm physiology.

Migration in the Vertebrate Host

Movement of the worms in the dermoepidermal region may continue for several days. There is evidence that the schistosomulae secrete a collagenaselike enzyme that may act on the epidermal basement membrane. However, the young worms appear to pass through sebaceous glands into the dermis rather than directly through the basement membrane. In the dermis, schistosomulae rather quickly enter lymphatic vessels or veins.

Some passive movement of schistosomulae in the venous or lymphatic system may now ensue, with the worms eventually reaching the pulmonary circulation. Virtually nothing is known of subsequent migration, insofar as specific responses of the worm are concerned. Essentially no growth of the worms occurs until the worms are found in the liver eight to ten days after infection of the mammal. In the liver, growth is quite dramatic, the worms doubling in size every three days in the third to fourth week and doubling every two days in the fourth to fifth week. During this latter period the worms migrate against the direction of blood flow into the mesenteric veins. It may be mentioned that the mesenteric venous system does not contain the valves found in some other parts of the venous system. Thus there is little to impede the migratory movements of the almost adult schistosomes. The growth and morphological changes in *S. mansoni* in the vertebrate are shown in Fig. 3–13.

Feeding of Adult Schistosomes

Schistosoma appears to have a functional digestive tract. The gut cells possess elongate protoplasmic processes suggestive of an absorptive surface. What has been termed hematin is commonly found in the gut of schistosomes, and there is reason to believe that the worms ingest erythrocytes. It has been reported that the addition of red cells to culture media improves the survival of schistosomes maintained in vitro. The fact that the worms show little growth until they have reached the portal circulation of the liver may suggest that the dissolved nutrients in the portal blood arriving from the host's digestive tract are of more nutritional significance than the erythrocytes. This is also suggested by the observation that when mice infected with *S. mansoni* were fed a low protein diet, the worms were stunted and produced fewer eggs than worms from control animals, although the low-protein diet did not affect the growth of the host. Red blood cells constitute a deficient food for a number of other organisms. *S. mansoni* adults contain an enzyme specifically hydrolyzing the globin moiety of hemoglobin. The relative significance of this enzyme in liberating amino acids, as compared with amino acids absorbed from blood plasma, cannot be quantitatively evaluated on the basis of information now available.

It has been demonstrated that the worms absorb nutrients through the tegument (Fig. 3–14). The structure of the tegument is similar to that of *Fasciola,* which is also known to have an absorptive function. Further, the localization of phosphatases in the tegument is consistent with digestive-absorptive function.

Metabolism of Adult Schistosomes

Carbohydrate appears to be the major energy source for schistosomes. *S. mansoni* metabolizes an amount of glucose equivalent to a fifth of its dry weight per hour. The worms contain considerable amounts of glycogen, 14.3 in males and 3.5 in females, as percentages of the dry weight. Energy metabolism is primarily anaerobic in character, since 80

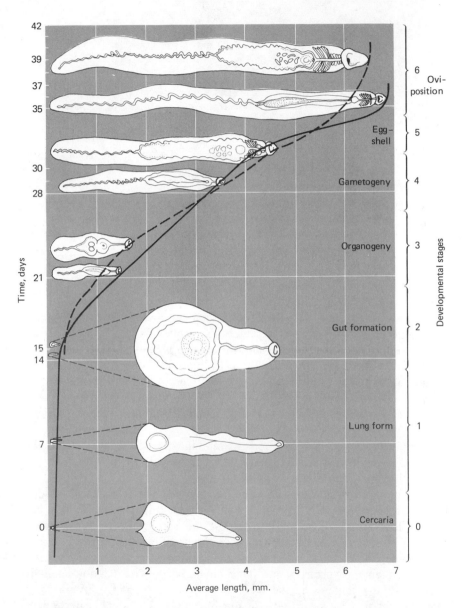

Fig. 3–13. Growth and morphological changes in *Schistosoma mansoni* in the vertebrate host. (After Clegg, 1965.)

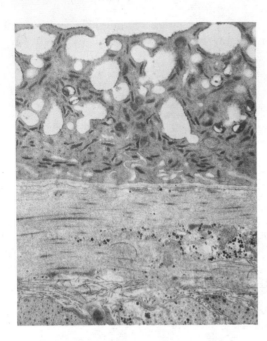

Fig. 3—14. The syncytial tegument of *Schistosoma mansoni.* Note the cryptic structure of this syncytial epidermis, which markedly increases surface area ($\times$ 15,800). Compare with the tegument of *Fasciola* shown in Figs. 3—8 and 3—9. (Courtesy of Dr. James Byram.)

to 90 percent of the carbohydrate metabolized is converted to DL-lactic acid and excreted into the surrounding medium.

The intermediate steps involved in the degradation of glucose to lactic acid by schistosomes seem to resemble the sequences in vertebrate tissues. However, the specific properties of schistosome glycolytic enzymes differ in some important respects. In addition to immunological and kinetic differences, phosphofructokinase, catalyzing the reaction

$$\text{Fructose-6-phosphate} + \text{ATP} \rightleftarrows \text{fructose-1,6-diphosphate} + \text{ADP},$$

is inhibited by trivalent antimony compounds, whereas the mammalian enzyme is unaffected. These antimony compounds have been important in the chemotherapy of schistosomiasis.

Although schistosomes consume oxygen when it is available, there is little evidence that it is of significance in energetic terms. It has been shown that certain cyanine dyes inhibit the respiration of schistosomes, both in vitro and in the animal host, without killing the worms. It is of significance, however, that this respiratory inhibition results in an inhibition of shelled embryo production. This might be due to a requirement for oxygen in the tanning of the embryo shell (see p. 58). It may seem paradoxical that a worm living in the blood has a negligible dependence on aerobic energy metabolism. This is discussed further on p. 169.

There is no evidence that adult schistosomes can metabolize lipids.

The worms have a specific acetylcholinesterase that is probably involved in neural function.

Only limited data are available on the metabolism of nitrogen compounds in *Schistosoma*. Several transaminases have been described in *S. japonicum* by Chinese researchers. In this species antimony compounds used in treating schistosomiasis were found to inhibit the reaction:

Glutamic acid + pyruvic acid → alanine + α-ketoglutaric acid.

However, host transaminases are not affected by antimony compounds. It would be of interest to determine whether transaminases of *S. mansoni* are antimony-sensitive, since as already remarked, a glycolytic enzyme of this species is known to be inhibited by antimony compounds.

Many attempts have been made to culture *S. mansoni* outside its vertebrate host. The greatest success has been obtained in chemically undefined media, but normal growth and reproduction of the worm have not been reported. Some observations suggest that physical conditions of cultivation may be of considerable importance. Therefore, although cultivation attempts have shed limited light on specific nutritional requirements of *S. mansoni,* it does not seem likely that the requirements will involve peculiarly exotic compounds.

REFERENCES

Dawes, B., and D. L. Hughes. 1964. Fascioliasis: The invasive stages of *Fasciola hepatica* in mammalian hosts. *Adv. in Parasitol.* 2:97.

Florkin, M., and B. T. Scheer (eds.). 1968. *Chemical Zoology,* Vol. 2. *Porifera, Coelenterata and Platyhelminthes.* New York: Academic Press, Inc.

Smyth, J. D. 1966. *Physiology of Trematodes.* San Francisco: W. H. Freeman and Co., Publishers.

Stirewalt, M. A. 1971. Penetration stimuli for schistosome cercariae. In *The Biology of Symbiosis.* T. C. Cheng (ed.). Baltimore: University Park Press.

As a class of flatworms the tapeworms differ sharply from the other members of the phylum. They are typically segmented (strobilate), with each segment containing a set of male and female reproductive organs. One end of this animal is differentiated as a scolex or holdfast organ. The tapeworms lack all vestiges of a digestive tract and may have evolved from gutless, free-living ancestors. Without exception the tapeworms are parasitic, and as reproductive adults they usually inhabit the small intestine in vertebrate hosts.

There is every evidence that the tapeworms were ancient parasites of fishes. They have evolved in concert with the vertebrates and now live in all groups of vertebrate hosts. Many of them show considerable host specificity, with a given worm species living in a limited number of closely related host species. As will become apparent, they show a number of interesting specializations for parasitism.

HYMENOLEPIS DIMINUTA

The rat tapeworm, *H. diminuta,* has an almost cosmopolitan distribution in the rats associated with man. It is of negligible significance to human health, although occasional infections of man have been reported. However, it is probably the best known of the tapeworms, having been the subject of numerous experimental investigations.

The Life Pattern of *Hymenolepis diminuta*

The strobilate adult *H. diminuta* may be up to 70 cm in length, and it lives in the small intestine of its rodent host. Each of the thousand or so segments contains male and female organs. As the shelled embryos

Tapeworm Parasites

chapter 4

(*oncospheres*) develop in the segment, the remaining organs tend to degenerate, and the segment is filled by the embryos. These segments detach from the strobila and undergo dissociation of the tissues, liberating all or part of the shelled embryos. These in turn pass out of the host in the feces in a stage that is fully infective for the intermediate host. If the shelled embryo is eaten by any one of a number of insects, particularly insects living in cereal foods, the oncosphere penetrates the insect gut and passes into the hemocoele. Here the worm undergoes growth and differentiation forming the *scolex* (head) characteristic of the strobilate adult (Fig. 4–1). The fully developed larval form, or *cysticercoid,* is enclosed in a small cyst and is infective for the rodent host. If the infected intermediate host is eaten by a rodent, the young worm excysts in the small gut. The free scolex attaches to the mucosa, and growth of a strobila ensues. Fully developed shelled embryos are present in the feces 17 days after infection of the rat. This life pattern is diagrammed in Fig. 4–2.

Infection of the Invertebrate

When the shelled embryo of *Hymenolepis* is eaten by the insect host, the outer stiff shell is mechanically ruptured by the mouthparts of the insect. The mechanism for liberation of the oncosphere from the shell differs in various cestode species. In *Taenia* the shell is removed by the action of host digestive secretions. The liberated six-hooked *Hymenolepis* embryo, or oncosphere, is thought to penetrate the gut wall of the insect by the combined mechanical action of the hooks and secretions from the so-called penetration glands (Fig. 4–3). The growth of the young worm in the hemocoele of the insect mainly involves the formation of the definitive scolex and its withdrawal into a cystic structure (Fig. 4–2). Completion of cysticercoid development requires 5 to 15 days, depending on the temperature, and the fully developed larval form remains in a state of arrested development for the life of the insect. Occasionally a developing cysticercoid may be infiltrated by blood cells of the insect host. When this occurs, the developing worm is killed. The vertebrate host becomes infected by eating an arthropod containing fully developed cysticercoids.

Excystment in the Vertebrate

The juvenile forms of most tapeworms are enclosed in cystic structures (Fig. 4–2). Since cestodes enter the vertebrate host by being eaten, usually in the body of the intermediate host, the juvenile worms are exposed to the acid environment of the stomach. The cystic structures protect the young worm during its passage through the stomach. If ex-

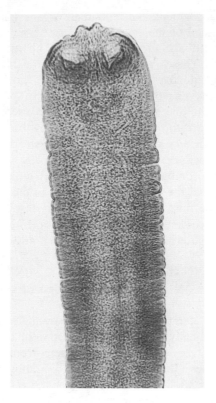

Fig. 4—1. The scolex, or head, of the rat tapeworm, *Hymenolepis diminuta.*

cysted worms are fed to the rat host, the worms do not survive passage through the stomach, although they develop quite well if introduced into the small intestine. Excystment occurs in the small intestine, and *H. diminuta* requires the combined effects of bile salts and trypsin to induce excystment. It is not known whether the young worm secretes any factor that might be involved in excystment, as appears to be the case in some trematodes and nematodes (see pp. 64 and 114). The efficiency of establishment in the rat is quite astonishing. A skillful laboratory worker can administer a single cysticercoid to a rat and almost invariably obtain a single fully developed worm.

Growth of the Strobila

H. diminuta grows quite rapidly in the vertebrate host, showing exponential increase in weight for about ten days. During this period, strobilization (segmentation) occurs. In looking at tapeworm growth, various workers have tended to overemphasize the formation of strobilar segments from a region behind the scolex and to neglect the fact that after a segment forms, it continues to grow in three dimensions. Growth in the

If *Hymenolepis* eggs are ingested by a suitable
intermediate host, they develop into infective cysticercoids.
Any one of a number of grain infesting insects can serve
as the intermediate host.

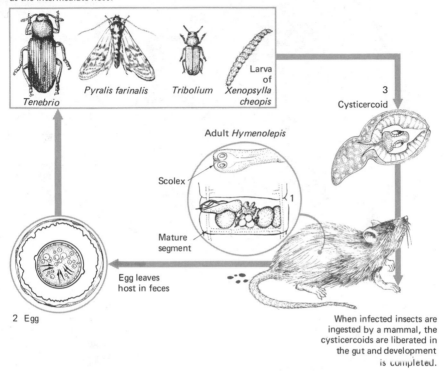

Fig. 4–2. The life pattern of *Hymenolepis diminuta*.

weight of the segment continues after its formation and is accompanied by
the differentiation and maturation of the structures of the genital complex.
Each segment contains a complete set of both male and female organs at
maturity.   Growth of the segment continues until the appearance of
shelled embryos about nine days after formation of the segment.   One interesting aspect of growth and maturation of the segment is the fact that
after a segment has been formed in a host on an adequate diet, its differentiation and maturation seem to be quite independent of segmental proliferation and growth.   For this reason the shedding of shelled zygotes
typically occurs 17 days after infection of the vertebrate, irrespective of the
growth in size of segment or strobila or the rate of segment formation.

A fully developed strobila living in an adequately fed host seems to
grow at an essentially constant daily rate, with egg-filled segments breaking down and passing out of the host at a rate about equal to the growth
rate.   Under such conditions the worm obviously tends to maintain a constant size over a period of time.   Except for accidents, the growing strobila

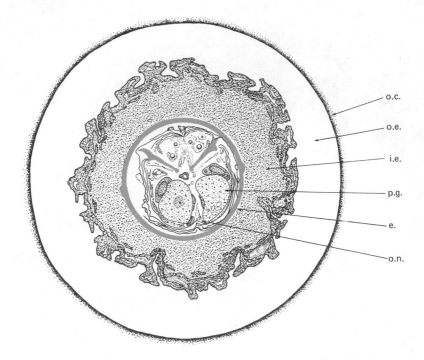

Fig. 4–3. The shelled oncosphere of *Hymenolepis diminuta*. Abbreviations: o.c., outer coat; o.e., outer envelope; o.n., oncosphere; p.g., penetration gland; i.e., inner envelope; e., embryophore. (After Pence, 1970.)

shows no signs of death, although the senescent events in the terminal segment tissues may be construed as an aging phenomenon. The worm will evidently live as long as its rodent host. As a matter of fact, it has been shown by transplanting from host to host that the worm can live for 14 years, and there is reason to believe that it might live for a much longer period of time.

If the strobila is broken in the neck region, a new strobila will regenerate. Schiller utilized this characteristic in studying the effects of X-irradiation on this organism. The strobila was cut and most of the worm's tissues retained as a control preparation. The scolex and neck were then irradiated and transplanted to a new host. After regeneration of a new strobila from the irradiated tissue, the major part of the worm was again removed for examination. By carrying this out repetitively Schiller could distinguish permanent changes from acute radiation effects of a nonpermanent character.

It should be emphasized that the preceding description of growth (p. 83) is the pattern seen in *H. diminuta* and that some tapeworms certainly differ. The closely related *Hymenolepis nana* and *H. citelli*, for example, show more general evidences of senescence and shorter life spans.

A word must be said concerning the growth of the *H. diminuta* stro-

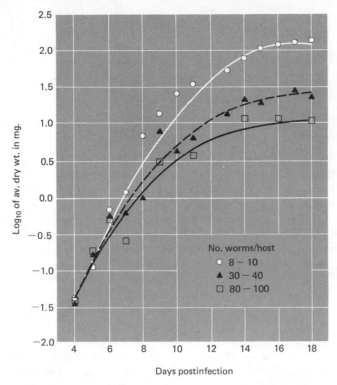

Fig. 4—4. The effect of intensity of infection of the rat host on the size of *Hymenolepis diminuta*. (After Roberts, 1961.)

bila as it is affected by the presence of other worms of the same species. The size of the individual worm (and its growth rate) is inversely related to the number of worms in the rat host (Fig. 4–4). The relationship is that of a rectangular hyperbola, and it has been hypothesized that it is related to the amount of available energy source (glucose) in the host gut. Since the absolute quantity of free glucose in the small intestine is a function of ingestion rate, stomach emptying rate, rate of hydrolysis of complex sugars, and rate of absorption by the mucosa, it may be postulated that the amount of free glucose available to the worm is growth-limiting and will only support the growth of an essentially fixed amount of tapeworm tissue, whether this is represented by 5 worms or by 25 worms.

Nutritional Mechanisms of *H. diminuta*

Since tapeworms do not have a digestive tract, food must be absorbed through the external surface. The tapeworm surface is a syncytium composed of extensions of cells lying below the outer layer. Many previous

Fig. 4–5. The syncytial tegument of *Hymenolepis diminuta.* Note the connection (c) of the microvillar system with a cell lying beneath the tegument. This connection penetrates the muscle layer (m) (8,200). (Courtesy of Dr. R. D. Lumsden.)

authors referred to this outer layer as a cuticle, but it is not a cuticle in the sense that biologists ordinarily use the term (Fig. 4–5). The surface syncytium has been shown to have a number of attributes of the intestinal mucosa of other kinds of organisms. For example, in *H. diminuta* there are systems for the mediated transport of a number of organic compounds, and these systems exhibit group specificity. Amino acids appear to be transported through four qualitatively different systems, although there is overlapping specificity among them. One of the best known of these systems in *Hymenolepis* is one involved in the uptake of diamino acids, such as arginine and lysine. Monoamino acids do not react with this system, except for histidine, which with its imidazole group may act like a diamino acid.

Since amino acids compete for absorption mechanisms and since the worm lives in an environment containing numerous amino acids rather than single ones, it is of significance that the absorption of a single amino acid from a mixture follows predictions derived from kinetic studies on pairs of amino acids. Thus, using the Michaelis-Menten formulation for enzyme kinetics, it was found that the rate of uptake of a single amino acid at a given concentration in a mixture is expressed by

$$v = \frac{V}{\dfrac{(K_t)}{(S)} + 1 + \dfrac{(K_t S^1)}{(K_i^1)\,(S)} + \dfrac{(K_t S^2)}{(K_i^2 S)} + \ldots + \dfrac{(K_b S^n)}{(K_i^n S)}}$$

where v = rate of uptake of amino acid at concentration S with an experimentally determined K_t (equivalent to Michaelis constant) and V (maximum rate). Inhibition constants of other amino acids (K_i) are determined experimentally, and concentration of other amino acids (S^1, S^2, etc.) are known.

It is important to point out that the free amino acids available to a worm in the vertebrate small intestine are not a simple function of the host's diet. It has been shown that the free amino acids in the gut arising from dietary sources are markedly diluted by amino acids from endogenous sources. Although the absolute concentrations of amino acids vary as a function of the dietary state, the relative concentrations of amino acids (ratio of one amino acid to another) remain relatively constant and independent of diet. The rate of absorption of an amino acid by an organism such as *Hymenolepis* will be to some extent determined by the other amino acids present, since the worm does not carry out digestion of proteins or peptides in the environment. If this interpretation is considered further in terms of the worm's lack of capacity for amino acid biosynthesis, the host mechanisms for maintaining amino acid homeostasis in the gut lumen would appear to be important in regulating the amino acid nutrition of *Hymenolepis*.

Very high rates of exchange of amino acids between the host and *Hymenolepis* have been demonstrated by measuring the movement of labeled amino acids from the host to the worm and from the worm to the host. When *Hymenolepis* containing labeled methionine or 1-aminocyclopentane-1-carboxylic acid (an amino acid not metabolized by rat or worm) is transplanted to a rat, 50 percent of the labeled amino acid passes out of the worm in six minutes (Fig. 4–6).

The hexose sugars, glucose and galactose, are transported through a common mechanism that does not appear to react with other classes of organic compounds, including hexose sugars of inappropriate configuration. On the other hand, purines and pyrimidines enter the tissues through specific mechanisms at low concentrations, but at higher concentrations diffusion appears to be the major mode by which these substances enter the worm. Fatty acids show a similar duality in the mechanisms for absorption. In this last group there are at least two qualitatively different mechanisms showing specificity as evidenced by the fact that short-chain volatile fatty acids compete with one another in their absorption and long-chain fatty acids compete with one another. However, short-chain fatty acids do not compete with the absorption of long-chain ones, nor conversely. Cholesterol is also absorbed by *H. diminuta,* but it has not been determined whether this occurs through a mediated process.

Fairbairn and his colleagues have shown that *Hymenolepis* is extremely limited in its capacity to synthesize fatty acids. There may be some chain-lengthening of higher fatty acids, but de novo synthesis has not

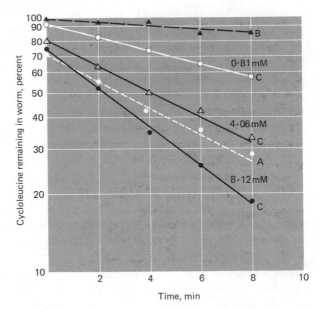

Fig. 4–6. The efflux of a ^{14}C-labeled amino acid from *Hymenolepis* in the rat gut (curve A), in Ringer's solution (curve B), and in an amino acid mixture of differing total amino acid concentration (curves C). (From Arme and Read, 1969.)

been demonstrable. Thus it would seem that the specific fatty-acid components of *Hymenolepis* tissues would be to a great extent dependent on what fatty acids are available in the host gut. Harrington observed that there were quantitative differences in the fatty-acid components of worms reared in hamsters compared to those of worms reared in rats, and Fairbairn and his collaborators showed that the relative quantities of fatty acids found in the lumen of the rat gut are to a considerable extent independent of dietary composition. In other words, like the amino acids, the fatty acids available to *Hymenolepis* are determined to a greater extent by the kind of host in which the worm occurs than by the diet of such a host.

We thus arrive at a concept of *Hymenolepis* as an organism living in an environment in which concentrations of major nutritional components are highly regulated by the host. The parasite may induce an increased flow of metabolites into the gut lumen, and this would represent a stimulus → reaction in which the host furnishes the basic regulatory mechanism. The carbohydrates represent an exception in that the free carbohydrate in the lumen of the intestine is mainly of dietary origin, and, as will be indicated below, this exception is manifested in the day-to-day sensitivity of *H. diminuta* to the carbohydrate ingested by the host.

Compared to the intestinal mucosa of some animals, the surface syncytium of *Hymenolepis* seems to have a limited digestive function. The worm does not appear to secrete proteolytic or amylolytic enzymes, and it is incapable of hydrolyzing protein or oligosaccharides. On the other hand, it hydrolyzes ester phosphates at the surface boundary. This has been demonstrated by taking advantage of the worm's impermeability to certain compounds. For example, the worm is quite impermeable to the sugar fructose. This sugar is not transported and thus is not metabolized by the worm. However, when the worm is incubated with fructose-1-phosphate, free fructose is liberated into the medium. Other phosphate esters compete in this hydrolysis. An ester such as glucose-6-phosphate is hydrolyzed, and the liberated glucose is transported into the tissues where it may be metabolized. What may be the functional significance of such surface hydrolysis? Two aspects suggest themselves. First, such a system would allow the entry of the products from hydrolysis of polar compounds to which the worm is not permeable; second, the entry of phosphate esters into a worm cell might interfere with the regulation of metabolism, and surface hydrolysis would prevent such entry. Interestingly, the worm does not appear to have the surface enzymes for the hydrolysis of glycosidic bonds such as those of maltose or sucrose. It is thus limited in its utilization of sugars to monosaccharides, since it is not capable of transporting disaccharides into the tissues.

In contrast to the sugars, *H. diminuta* can apparently hydrolyze monoglycerides, and recent studies suggest that the hydrolysis occurs on the surface of the worm. Although it has been suggested that tapeworms may take up material from the environment by pinocytosis, there is no evidence that this occurs. Recent studies by Lumsden and his colleagues indicate that pinocytosis does not occur in *Hymenolepis,* and other studies have shown that this worm does not take up some rather small molecules, such as fructose.

Host Dietary Components

As indicated, carbohydrate is a major constituent of the host diet to which *H. diminuta* and some other tapeworms are sensitive. The worm requires carbohydrate for growth, and reproduction and growth rate are functions of carbohydrate quantity at low dietary concentrations. Further, the quality of carbohydrate in the host diet is important. Since the worm is incapable of hydrolyzing oligosaccharides and is only capable of metabolizing glucose or galactose, the quality of host dietary carbohydrate is significant in determining worm growth. Some recent work suggests that even galactose is a poor substrate for growth, and glucose may prove to be the only satisfactory energy source for this organism.

Although Chandler reported that *H. diminuta* was independent of vitamins in the host diet in normal animals, Roberts and his colleagues have recently shown that if coprophagy is prevented in the rat, the worm is deleteriously affected by omission of some water-soluble vitamins from the host diet.

Cultivation in Vitro

The worm *H. diminuta* was successfully grown in vitro by Berntzen in 1961. This provides an interesting illustration of the importance of an open mind in science. Over a number of years, various investigators had attempted the cultivation of cyclophyllidean tapeworms, and the results had been rather discouraging. However, an American undergraduate, Berntzen, undismayed by the failures of others, undertook the cultivation of *H. diminuta* in a flow culture system containing a complex undefined medium. He was successful and the worm grew. Subsequently Berntzen's methods have been applied with good results in growing a number of other helminths outside their hosts.

More recently Schiller developed a stationary culture method for *H. diminuta,* and the closely related *Hymenolepis nana* has since been grown in roller tubes. All of these culture methods are important technological advances, allowing new uses of these organisms in developmental biology and physiology.

Metabolism of the Strobila

A number of investigations have been made of energy metabolism in *H. diminuta.* Although the worm takes up oxygen at a low rate under aerobic conditions, there is evidence that the transfer of electrons to oxygen is of negligible energetic significance in the economy of the organism. There is neither enhancement of glycogenesis nor change in rate of glucose consumption in the presence of oxygen. Metabolism is fermentative in character, with succinic, lactic, and acetic acids excreted into the ambient medium. Carbon dioxide fixation appears to be required in energy metabolism and is known to be required for succinate formation. In the absence of carbon dioxide, there is a sharp depression of glucose utilization and of glycogenesis and a decrease in the level of ATP in the tissues.

Glycogen is present at very high concentrations in the tissues of *H. diminuta,* sometimes representing more than 50 percent of the dry weight. When the host is deprived of carbohydrate in the diet, the glycogen content of the worm falls rapidly. When such starved worms are given glucose in vitro, net glycogenesis also proceeds rapidly (Fig. 4–7). Bueding and his colleagues have studied the glycogens of *H. diminuta* and of

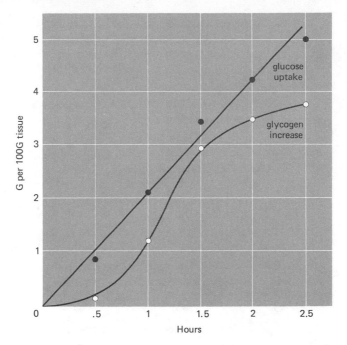

Fig. 4–7. Glycogenesis in *Hymenolepis* removed from previously starved hosts and incubated with glucose. (After Read, 1967.)

some other helminths. When the glycogen is removed by mild water extraction and examined in the analytical centrifuge, two major peaks of molecular weight are discerned (Fig. 4–8). Upon starvation these molecular species are utilized at different rates, and during glycogenesis the resynthesis of different molecular species occurs at different rates.

The major initial pathway for carbohydrate degradation seems to be the Embden-Meyerhof sequence of reactions. The pentose phosphate shunt is present but is not utilized to a major extent for glucose degradation. Fixation of carbon dioxide may occur at the phosphoenol pyruvic acid stage or may involve the so-called "malic enzyme," both systems being present in the worm.

Thus the general pattern of energy metabolism in *H. diminuta* is that of an anaerobic organism that excretes partially oxidized organic compounds into its environment. It does not appear to have a complete tricarboxylic acid cycle for the degradation of glucose to carbon dioxide and water.

H. diminuta is extremely limited in its ability to synthesize amino acids. It has an unusually simple capacity for performing transaminations and may be capable of producing only alanine, glutamic acid, and aspartic acid. Further, it does not seem to have any capacity for the utilization of

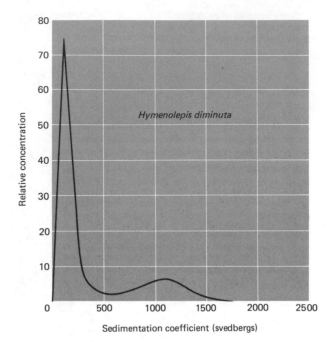

Fig. 4–8. Profile of molecular species of glycogen in *Hymenolepis diminuta*, as determined in the analytical centrifuge. (After Orrell, Bueding, and Reissig.)

amino acids as energy sources. In studying the incorporation of lysine into *H. diminuta* protein, Harris and Read showed that protein synthesis is driven by glucose metabolism. Starved worms required glucose and carbon dioxide for the incorporation of significant quantities of lysine. It was remarked that *H. diminuta* has a metabolism resembling that pictured in some elementary textbooks, a metabolism in which catabolism and anabolism are linked mainly at the energy transfer level, with very little chemical interchange between degradative and synthetic processes.

Movement in the Vertebrate Host

Although it had long been assumed that *Hymenolepis* maintains a more or less stable position in the host intestine after reaching its definitive size, recent studies have shown that the worm undergoes a diurnal migration along the length of the small intestine (Fig. 4–9). This is related to the feeding pattern of the host as shown by the finding that if the host is forced to eat in the day hours rather than the night, the pattern of migration is correspondingly modified. It is not known whether the worm responds to some component of the host diet or to secretions associated

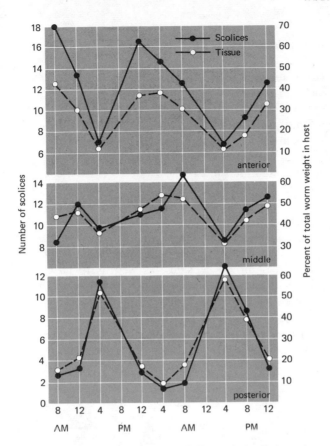

Fig. 4–9. The diurnal migration of *Hymenolepis diminuta* in the small intestine of the rat host. Animals were infected with 30 worms. "Anterior" refers to the first 10 inches, "middle" to the second 10 inches, and "posterior" to remainder of the small intestine. (After Read and Kilejian, 1969.)

with the feeding pattern of the host. It remains to be determined whether other tapeworms show such patterns of movement in the vertebrate host. Those species that, unlike *H. diminuta,* have hooks embedded in the host mucosa may not so readily move the scolex, although the body of the strobila may show diurnal linear movement in the host gut.

ECHINOCOCCUS GRANULOSUS

This tapeworm has a cosmopolitan distribution but is found most frequently in the temperate zones of the world. Its larval stages produce serious disease in man and various herbivorous animals.

The Life Pattern of *Echinococcus granulosus*

The strobilate adult of *Echinococcus granulosus* is a small worm, 3 to 6 mm in length, that lives in the small intestine of dogs, wolves, foxes, and related carnivores (Fig. 4–10). Single gravid segments are shed by the worm and disintegrate in the host's digestive tract, liberating the shelled embryos. When the shelled embryos are then ingested by sheep, cattle, pigs, or other ungulates or by man, the oncosphere embryos emerge from the shells in the small intestine. The embryos enter the wall of the host's digestive tract and typically are carried in the blood to the liver or the lungs. In the viscera, each young worm becomes vacuolated and slowly grows into a cystic structure (see Fig. 4–11). This cyst consists of two well-defined layers, an outer laminated wall and an inner germinative membrane (Fig. 4–12). The inner germinative membrane gives rise to multiple daughter cysts or brood capsules, as well as to larval scolices called *protoscolices*. The brood capsules in turn give rise to numerous protoscolices. Growth of the main larval cyst, or *hydatid*, is slow, but the organism may reach a diameter of 15 cm in 10 to 12 years, depending on its location in the host. As might be anticipated, completion of the life cycle depends on the ingestion of protoscolices by a carnivorous host, with

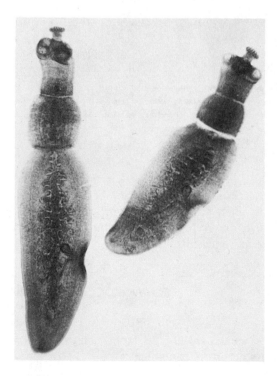

Fig. 4–10. The adult of *Echinococcus granulosus*.

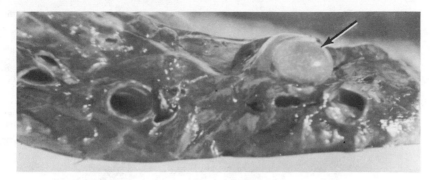

Fig. 4—11. A cyst of *Echinococcus* in the lung of a deer.

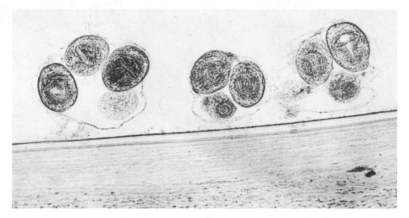

Fig. 4—12. The cyst wall of *Echinococcus granulosus*. Note daughter cysts developing on inner germinative membrane.

the ensuing growth of a strobila. The life pattern is summarized in Fig. 4–13.

Development of the Hydatid

When the embryo of *Echinococcus* passes out of the host in the feces, it is enclosed in a rather thick, striated shell known as the embryophore. The oncosphere embryo is also covered by a second thin oncosphere membrane (Fig. 4–14). If the embryo is ingested by a mammal, the embryophore is not affected by the pepsin in the stomach but is acted upon by pancreatic enzymes, particularly trypsin, with the resulting liberation of the oncospheres. Activation of the oncospheres occurs in human bile, intestinal juice from cattle or sheep, or in artificial pancreatic juice. Onco-

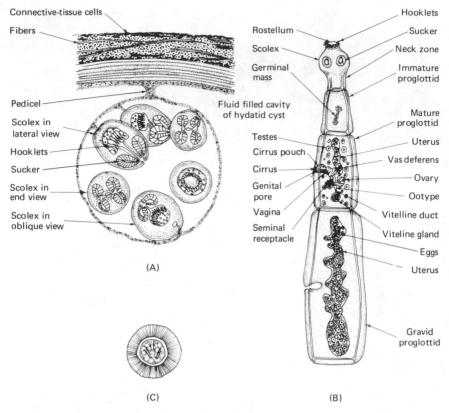

Connective-tissue cells
Fibers
Pedicel
Scolex in lateral view
Hooklets
Sucker
Scolex in end view
Scolex in oblique view

(A)

Rostellum
Scolex
Germinal mass
Fluid filled cavity of hydatid cyst
Testes
Cirrus pouch
Cirrus
Genital pore
Vagina
Seminal receptacle

Hooklets
Sucker
Neck zone
Immature proglottid
Mature proglottid
Uterus
Vas deferens
Ovary
Ootype
Vitelline duct
Viteline gland
Eggs
Uterus
Gravid proglottid

(C)   (B)

Fig. 4—13. The life pattern of *Echinococcus granulosus*. (A) Portion of hydatid cyst from the herbivorous intermediate host. (B) Adult from the intestine of the carnivorous host. (C) Shelled embryo voided from the carnivorous host.

spheres are immobilized if treated with dog bile or with jejunal juice from cats or dogs. This latter observation is consistent with the fact that the larval stages do not readily infect dogs or cats and is the converse of the sensitivity of protoscolices of bile salts from herbivorous animals (p. 98).

Although little seems to be known of the mechanism by which the freed oncospheres of *Echinococcus* penetrate the mucosa of the intermediate host, it is thought that they pass to the liver and lung by way of the vascular system. In susceptible animals, foci of infection can be found in the liver within a day or so after ingestion of shelled oncospheres. Unilocular vesiculation of the young worm begins after three or four days, and by ten days the germinal membrane of the young cyst is recognizable. The laminated outer membrane then begins to develop. In some hosts there may be invasion of the parasite without establishment. Host tissue reaction may be severe, and the parasite may be quickly walled off by

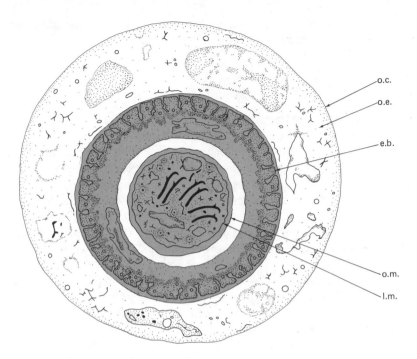

Fig. 4–14. The shelled oncosphere typical of taeniid cestodes such as *Echinococcus.* Abbreviations: o.c., outer coat; o.e., outer envelope; e.b., embryophore blocks; o.m., outer oncosphere membrane; l.m., limiting membrane of oncosphere. (After Morseth, 1968.)

fibrous connective tissue, forming what is termed a *pseudotubercle.* If the host reaction is less violent, the formation of an outer fibrous connective tissue envelope may only inhibit development. In such instances a cyst that fails to produce protoscolices may develop.

In those hosts in which rapid development of fertile cysts occurs, there is some host reaction, with the formation of a capsule of host connective tissue outside the laminated layer of the hydatid. The rate and extent of reaction by the host seem to be dependent on the strain of both the host and the parasite. The germinal membrane of the developing hydatid begins to show local accumulations of nuclei and projections of cellular aggregations into the lumen of the cyst that is filled with fluid. These aggregations form protoscolices that in turn vacuolate and produce daughter cysts, commonly called brood capsules. Each brood capsule has its own germinal membrane and gives rise to protoscolices on this inner layer of the capsule. Thus a single oncosphere may give rise to thousands of new individuals, as each protoscolex is potentially capable of developing into an adult worm.

The hydatid may live in the intermediate host for very long periods of time (up to 50 years). However, older cysts tend to contain many dead protoscolices, which may be free in the fluid of the cyst.

Growth in the Definitive Host

When brood capsules containing protoscolices are eaten by a carnivore, the protoscolices are liberated through the action of pepsin in the stomach. However, the protoscolices do not evaginate in acid solution. In neutral solution the protoscolices will evaginate slowly. The addition of salts of bile acids markedly potentiates this evagination, and it seems probable that bile salts produce such an effect in the host. Interestingly, deoxycholate has a very deleterious effect on the protoscolices of *Echinococcus,* causing lysis of the tegument. This bile salt characteristically occurs in the bile of herbivorous animals that are unsuitable hosts for adult *Echinococcus.* Glycocholate, another bile salt of herbivores, is also toxic to *Echinococcus,* whereas taurocholate, a major component of the bile of carnivores, is tolerated by the cestode. Thus the nature of the bile salts in a host may be an important determinant of whether or not a given animal will serve as a definitive host.

The evaginated protoscolices become established in the small intestine, with the rostellum buried in a crypt of Lieberkuhn, and for the first four to five days in the host they undergo little growth. By the 14th day the scolex has grown in size, and the first segment has appeared. The genital anlage and a second segment appear by the 18th day. Growth continues and fully developed shelled embryos are formed by about the 35th day. The events of strobilar development are illustrated in Fig. 4–15. In very heavy infections in dogs, development may be markedly delayed, with no development of shelled embryos after as long as 60 days.

Growth in Vitro

A number of workers found that when protoscolices were cultured in various complex media, the worms formed small hydatid vesicles with a typical laminated layer on the outside. This of course is what happens when protoscolices are liberated in an extraintestinal site in a vertebrate. Smyth and his colleagues studied this phenomenon in some detail and found that *E. granulosus* protoscolices failed to differentiate a strobila in vitro in any number of liquid overlayers. If, however, protoscolices were cultured in vessels containing a coagulated serum base overlain with a complex liquid medium, the worms underwent segmentation, and the male and female genitalia developed to maturity.

It seems clear that for differentiation of a strobila, the worm requires

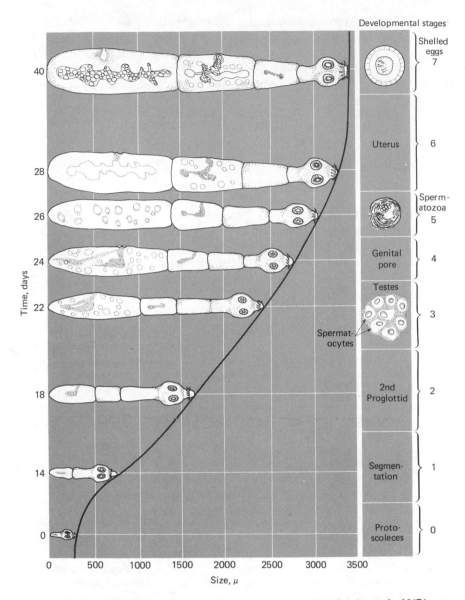

Fig. 4–15. The strobilar development of *Echinococcus*. (After Smyth, et al., 1967.)

a solid substratum. However, the results obtained with an agar base suggest that it is not only a simple matter of the physical character of the substratum, but that the chemical nature of the semisolid base is also important. It has been suggested that it is a matter of a requirement for protein, but there are other possibilities. Smyth showed that the portion of the protoscolex that is imbedded in the intestinal crypts is covered with the microvilli that are typical of the adult cestode tegument, whereas the region behind the suckers of the protoscolex does not possess fully developed microvilli. It has been postulated that the contact of this microvillar surface with a semisolid substrate is involved as a stimulus for release of a strobilization organizer, perhaps mediated through a neurosecretory mechanism. Further study should lead to the definition of the nature of the differentiation stimulus in *Echinococcus*.

Energy Metabolism of *Echinococcus*

Like many other animal parasites, the protoscolices of *Echinococcus* carry out aerobic fermentation of carbohydrate. The protoscolices excrete pyruvic, acetic, succinic, and lactic acids, and small quantities of ethyl alcohol. In the absence of oxygen there is an increased production of succinate, and no pyruvate is excreted. Under all incubation conditions, lactate is a major product, accounting for at least 50 percent of the degraded carbohydrate.

Echinococcus has at least four hexokinases specifically catalyzing phosphorylation of glucose, fructose, mannose, and glucosamine. As mentioned on p. 78, the trematode *Schistosoma mansoni* also possesses four hexokinases. However, there are readily demonstrable differences between the hexokinases of *Echinococcus* and those of *Schistosoma*. In addition to kinetic differences, schistosomes phosphorylate 2-deoxyglucose, but the tapeworms cannot. The fructokinase of *Echinococcus* is inhibited by glucose-6-phosphate, whereas the schistosome enzyme is not so inhibited.

There is evidence for the hexose monophosphate shunt in *Echinococcus*. Ribulose, ribose, sedoheptulose, and glyceraldehyde have been found in hydatid cyst fluid, and the following enzymes have been demonstrated in the tissues of the worm: glucose-6-phosphate dehydrogenase, 6-phosphogluconic acid dehydrogenase, transketolase, transaldolase, phosphopentose isomerase, ribokinase, 3-phosphoglyceraldehyde dehydrogenase, triose phosphate isomerase, and possibly phosphoketopentose epimerase. Cell-free preparations of *Echinococcus* produce appreciable amounts of pyruvate and lactate from ribose phosphate, but the data available suggest that the Embden-Meyerhof pathway, rather than the pentose shunt, is the major route for the production of succinate by the organism.

Carbon dioxide fixation in *Echinococcus* may occur through at least four pathways that have been demonstrated in the protoscolices. These include:

1. Phosphoenolpyruvate carboxykinase catalyzing

$$PEP + CO_2 + GDP \rightleftarrows \text{oxaloacetate} + GTP$$

2. Phosphoenolpyruvate carboxylase catalyzing

$$PEP + CO_2 \rightarrow \text{oxaloacetate} + P_i$$

3. Pyruvate carboxylase catalyzing

$$\text{Pyruvate} + CO_2 + CTP \rightarrow \text{oxaloacetate} + CDP + P_i$$

4. "Malic enzyme" catalyzing

$$\text{Pyruvate} + CO_2 + NADPH \rightleftarrows \text{malate} + NADP.$$

Agosin and Repetto showed that the major pathway for carbon dioxide fixation in the formation of succinate includes the malic enzyme. It may be pointed out that *Echinococcus* seems to differ from *Hymenolepis* in that, in the latter species, carbon dioxide fixation leading to succinate formation seems mainly to involve PEP carboxykinase (p. 91).

All of the intermediates of the tricarboxylic acid cycle are oxidized by *Echinococcus,* and intermediates of the cycle are labeled when the worm is incubated in the presence of $^{14}CO_2$. Further, the worm oxidizes acetate, glutamate, glyoxylate, and glycollate and contains pyruvic decarboxylase and acetyl-CoA-kinase. When *Echinococcus* protoscolices are incubated with $^{14}CO_2$ for 24 hours, worm protein contains labeled glutamic and aspartic acid. All of these findings are consistent with the view that this worm has a functional tricarboxylic acid cycle.

Echinococcus protoscolices take up oxygen when it is available. At tensions below about 0.8 mm^3, oxygen consumption rapidly falls to zero. Above 0.8 mm^3, the rate of oxygen consumption is linear with respect to oxygen tension up to about 3 mm^3. The oxygen tension has been measured in hydatid cysts and ranges from 1.25 to 2.25 mm^3 (Fig. 4–16). It thus appears that oxygen is consumed by this parasite in vivo. Further, Smyth and his colleagues have shown that oxygen is required for normal development of the strobila in vitro. However, *Echinococcus* protoscolices do not exhibit a Pasteur effect; the amounts of carbohydrate metabolized under aerobic or anaerobic conditions are not significantly different. This suggests that the worm may derive little benefit in terms of energy from the utilization of oxygen. As a matter of fact, the oxygen consumed would account for the complete oxidation of less than half of the carbohydrate metabolized.

Polysaccharides containing galactosamine, glucosamine, galactose, and glucose have been identified in the laminated membrane of hydatid cysts and protoscolices. One of these has an infrared spectrum identical with glycogen, whereas another (alkali-labile) has a mucopolysaccharide or a polysaccharide-protein structure. This is of special interest because of reports that an antigenic mucopolysaccharide that is seemingly identical with human P blood group substance occurs in the laminated membrane of *Echinococcus* cysts. This antigenic material is derived from the para-

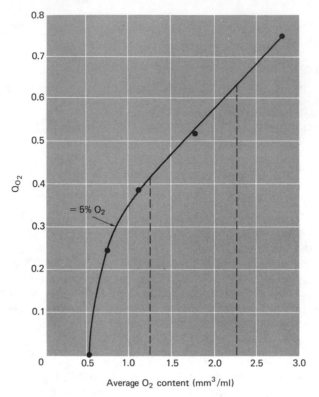

Fig. 4–16. Oxygen consumption of *Echinococcus* protoscolices as a function of oxygen tension. (From Read and Simmons, 1963.)

site rather than from the host, since it appears with the laminated membrane secreted by protoscolices incubated in vitro under conditions promoting vesicular differentiation rather than strobilization.

The Species of *Echinococcus*

Speciation in this genus is a complex and confusing state of affairs. Until the early 1950s, most parasitologists considered that there was a single species in the genus, *E. granulosus*. Different patterns of development of the hydatid were recognized, but it was thought that these variations could be attributed to development in different species of intermediate hosts. Rausch and Schiller in North America and Vogel in Europe reinvestigated this problem. Four types of hydatid have been distinguished. These are:

1. *Unilocular:* Characterized by single, well-defined bladders, in which the laminated membrane continuously encloses the germinal membrane.

2. *Multivesicular:* Characterized by many adjoining and connected bladders, each having its own separate germinal membrane.

3. *Alveolar:* Characterized by a malignant type of growth with jelly-filled proliferating vesicles embedded in a common dense stroma. The very thin laminated membrane does not restrict the germinal membrane that grows out into surrounding host tissues.

4. *Multilocular:* Characterized by many small bladders embedded in a common enclosing membrane. Some authors regard this as a stage in the development of the alveolar type of hydatid.

Careful study has shown that two species can be distinguished with certainty. These are *E. granulosus* and *E. multilocularis.* The two species differ remarkably in the patterns of larval development, in the morphology of the adult worm (Fig. 4–17), and in the hosts involved in the life cycles. The hydatid of *E. granulosus* develops in ruminants, as well as in

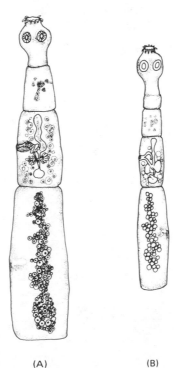

Fig. 4–17. The adults of *Echinococcus granulosus* (A) and *E. multilocularis* (B).

(A)   (B)

camels, horses, wallabies, pigs, and man; whereas *E. multilocularis* typically develops in rodents, insectivores, and man. Several subspecies of each of these forms have been described, mainly on the basis of distribution in isolated hosts. Another species, *E. oligarthrus,* has been described from South American felines, and *E. patagonicus* has been described from the Magellan fox in Patagonia. These two latter species are not well known.

Variation with rapid proliferation of *Echinococcus* mutant populations in new hosts would be predicted on a theoretical basis. The adult worm is hermaphroditic, and this would allow rapid segregation of recessive characters. Further, the development of the larval stages by polyembryony, with the production of thousands of new individuals having the same genetic makeup, could allow the rapid buildup of a mutant population in a new host or hosts.

REFERENCES

Florkin, M., and B. T. Scheer (eds.). 1968. *Chemical Zoology,* Vol. 2. *Porifera, Coelenterata and Platyhelminthes.* New York: Academic Press, Inc.

Read, C. P., and J. E. Simmons, Jr. 1963. The physiology and biochemistry of tapeworms. *Physiol. Rev.* 43:263.

Smyth, J. D. 1969. *The Physiology of Cestodes.* San Francisco: W. H. Freeman and Co., Publishers.

In some ways the Acanthocephala, or thorny-headed worms, are the most mysterious of animal parasites. All members of the phylum are parasitic. They are pseudocoelomate animals with separate sexes and, like the tapeworms, lack a digestive tract. Their evolutionary origins are quite obscure, and they show no close relationship to any other animal phylum. Adult acanthocephalans parasitize all major vertebrate groups, and arthropods are commonly utilized as intermediate hosts. Since the Acanthocephala are of little veterinary or medical significance, they have not received the attention that researchers have lavished on the protozoans, flatworms, and nematodes. Most of the experimental research on acanthocephalans has been carried out with the species *Moniliformis dubius,* which is readily maintained in the laboratory rat, and *M. dubius* will be used as an example for discussion.

The Life Pattern of *Moniliformis*

The adult male and female worms live in the upper small intestine of the rodent host. The shelled embryos (*acanthors*) are deposited in the gut and leave the host in the feces. If the acanthor is ingested by a cockroach, especially *Periplaneta americana,* it hatches from the shell and penetrates the wall of the insect's digestive tract. It passes through the gut wall and enters the hemocoele. Here it undergoes growth and differentiation, transforming into the *acanthella.* The fully formed *cystacanth,* a form that contains most of the structures of the adult worm, is infective for the vertebrate host, and if the roach is eaten by a rat, the worm everts the proboscis, which is embedded in the mucosa of the host's small intestine. Male and female worms grow to sexual maturity and copulate, and acanthors are de-

Acanthocephala

chapter 5

veloped in the pseudocoel of the female worm. The events of the life pattern are shown in Fig. 5–1.

Development in the Cockroach

In vitro, the shelled acanthor of *M. dubius* (Fig. 5–1) is stimulated by pH, and CO_2, and possibly by tonicity of the external medium, to secrete a chitinase that acts upon the chitin layer of the shell. It is presumed that this happens in the gut of the cockroach. The shell may also be subjected to the grinding action of the insect proventriculus, but a mechanical aid to hatching has not been directly demonstrated. In vitro, the hatched acanthor alternately retracts and extrudes the hooks, suggesting that these may function when it is penetrating the insect gut.

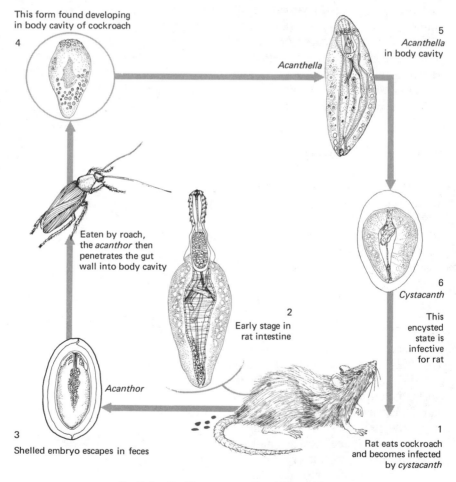

This form found developing in body cavity of cockroach

4

5
Acanthella in body cavity

Acanthella

Eaten by roach, the *acanthor* then penetrates the gut wall into body cavity

2
Early stage in rat intestine

6
Cystacanth

This encysted state is infective for rat

Acanthor

3
Shelled embryo escapes in feces

1
Rat eats cockroach and becomes infected by *cystacanth*

Fig. 5–1. The life pattern of *Moniliformis dubius.*

After entering the insect hemocoele, the acanthor becomes invested by host hemocytes, which lay down a mass of vesicular protoplasmic filaments around the developing parasite. The embryo becomes elongate and shows rapid differentiation of the proboscis, proboscis receptable, lemnisci, and other organs that make up the acanthella (Fig. 5–1). During the process the pseudocoelom forms by delamination of cells destined for differentiation as muscle cells of the body wall.

In the fully developed acanthella the proboscis is everted as in Fig. 5–1. The cystacanth is formed by the inversion of the proboscis and consolidation of the outer capsule. No further development occurs in the cockroach, and the cystacanth is a resting stage that is infective for the rat (Fig. 5–1).

Development in the Rat

When the dormant cystacanth is ingested by a rat, the juvenile worm is activated by bile salts and by the CO_2-bicarbonate system in the small intestine. The outer capsule of the cystacanth is acted upon by pepsin, but this is not essential for establishment in the host. Activation results in eversion of the proboscis, and the young worm becomes attached to the host mucosa.

Growth of the worm proceeds, with the differentiation of the muscle system of the body wall, followed by spermatogenesis and oogenesis. Embryos are produced after about five to six weeks of development in the rat, during which period female worms grow about 300-fold in length.

Nutrition of *Moniliformis*

Since the acanthocephalans lack a digestive tract, food is presumably taken in through the outer surface. The structure of the tegument is therefore of particular interest. Its general ultrastructure is shown in Fig. 5–2. As in the tapeworms the tegument is a syncytial structure. However, it differs sharply from the cestode tegument in a number of respects. The surface is not covered with microvilli but is ramified by blind canals leading inward. There is a well-developed outer fuzzy coat of mucopolysaccharide. A plasma membrane lies immediately beneath the fuzzy coat, and this limiting membrane appears to line the canal system completely. Some early reports suggested that the surface was capable of carrying out pinocytosis, but it now appears that the so-called pinocytotic vesicles are merely sections of the individual blind canals of the ramified surface.

Little seems to be known of the possible digestive function of the acanthocephalan integument. It is suggestive that phosphatase and leucine

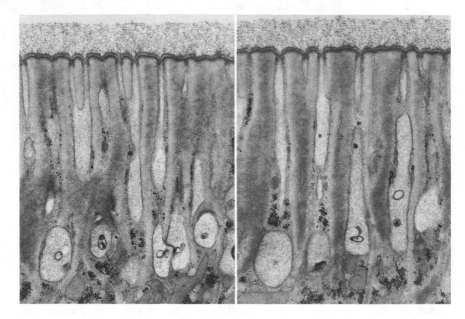

Fig. 5—2. The ultrastructure of two successive sections of the outer tegument of *Moniliformis dubius* ($\times$ 6,300). (Courtesy of Dr. James Byram.)

aminopeptidase have been demonstrated by histochemical methods in the tegumentary canals. *Moniliformis* metabolizes maltose, but it is not known whether this is hydrolyzed at the outer surface, as occurs in the vertebrate mucosa.

Moniliformis is dependent on the host diet as a source of carbohydrate. Worm glycogen falls sharply if the host is deprived of carbohydrate, and there is a diurnal fluctuation in worm glycogen that is correlated with the feeding pattern of the host. The worm is dependent on host dietary carbohydrate for growth. Worm growth is arrested if the host is placed on a carbohydrate-free diet for a few days during the initial period of worm growth, but growth commences again when carbohydrate is added to the host diet. No other data on nutritional requirements seem to be available.

Low molecular weight organic compounds are absorbed by *Moniliformis*. It is not known whether sugar enters by mediated mechanisms, but it may be suspected that such is the case. There is good evidence that the transport of amino acids occurs through mediated processes at concentrations resembling those in the digestive tract of the host. *Moniliformis* absorbs methionine against a concentration difference, and the rate of absorption is nonlinear with respect to concentration of methionine. Further, methionine absorption is competitively inhibited by leucine, isoleucine, serine, alanine, and valine, and reciprocal inhibitions have been observed. When mixtures of amino acids were tested as inhibitors of methionine absorption, they behaved as did mixtures tested with the tapeworm *Hymeno-lepis* (p. 86). At higher concentrations of amino acids some may enter the worm by diffusion, since the absorption rate does not reach a stable saturated level.

There have been only a few attempts to grow acanthocephalans in vitro, and these have been singularly unsuccessful.

Energy Metabolism of *Moniliformis*

The major storage carbohydrate in *Moniliformis* is glycogen, but the worm also contains considerable amounts of the disaccharide trehalose. Glycogen is readily synthesized from glucose, fructose, mannose, or mal-tose by intact worms. Trehalose is synthesized from glucose in homog-enate preparations fortified with UDPG and ATP, suggesting that the mechanism is similar to that reported to occur in insects and yeast.

Degradation of carbohydrate seems to occur through the Embden-Meyerhof scheme, several of the enzymes in this sequence having been demonstrated. Under aerobic conditions *Moniliformis* produces mainly formic and acetic acids and small amounts of lactic acid, as well as carbon dioxide. Anaerobically it produces significant quantities of succinic acid. Body wall preparations incorporate $^{14}CO_2$ into malate, succinate, and aspartate, and to a lesser extent, into fumarate, lactate, and alanine. When body wall preparations are incubated with ^{14}C-glucose, the above inter-mediates, as well as serine, are labeled. Interestingly, it has been reported that CO_2 is not required for normal rates of glycogenesis, as has been observed with the tapeworm *Hymenolepis*. This may indicate that the fixa-tion of CO_2 is less significant in the energetics of this worm than in the tapeworm.

The lack of tricarboxylic acid labeling when the worm is incubated with ^{14}C-glucose strongly suggests that a Krebs cycle is not present.

Resistance and Competition

The vertebrate host does not seem to develop resistance to *Monili-formis*. Thus, although the proboscis of the worm is embedded in the mucosa, there seems to be no molecular contact between host and parasite that would result in an immune response. There is some evidence that crowding may affect the size attained by individual worms. In heavy in-fections, worms are stunted. Further, when *Hymenolepis diminuta* is present as a second parasite in the rat, *Moniliformis* has a more restricted distribution in the gut and grows more slowly. Holmes suggested that this effect was due to competition between the two species of parasites for

carbohydrate energy sources. Little is known about such competitions between species of parasites, and the subject should be examined further. In
nature, infection of a vertebrate with more than one parasite species is the
rule rather than the exception.

REFERENCES

Crompton, D. W. T. 1970. *An Ecological Approach to Acanthocephalan Physiology*. London and New York: Cambridge University Press.

Florkin, M., and B. T. Scheer (eds.). 1969. *Chemical Zoology*, Vol. 3. *Echinodermata, Nematoda, and Acanthocephala*. New York: Academic Press, Inc.

Nicholas, W. L. 1967. The biology of the Acanthocephala. *Adv. in Parasitol.* 5:205.

The nematodes are usually considered to be a class of the phylum Aschelminthes. They occupy virtually every kind of habitat on the earth, with the exception that none of them have learned how to fly. In total number of species the nematodes probably outdo the insects, although only a small fraction of them have been described and named. Several thousand species are known to be parasites of plants and animals. The nematodes show a deceptive simplicity and consistency in their general structure. Modifications of structure superficially appear to be less extensive than those we see among the insects. On the other hand, the apparent consistency in body form may mask marked variations in physiology at the cellular level. This should be kept in mind in considering the forms to be discussed. Arthur and Sanborn (1969) regarded nematodes as an evolutionary "dead end." Rather they should be considered as evolutionary "radicals," since they occupy an astonishing array of ecological niches with minimal morphological change. They are the multicellular rivals of the bacteria.

HAEMONCHUS CONTORTUS

The barber-pole nematode, *Haemonchus contortus,* is considered to be the commonest and most pathogenic nematode parasite of sheep. It has a cosmopolitan distribution but may be undergoing rapid speciation under our very noses. It is of considerable economic importance in the sheep-raising areas of the world from Canada to Australia, and large amounts of time and money have been expended in attempting to work out satisfactory methods of control.

Nematode
Parasites

chapter 6

The Life Pattern of *Haemonchus*

The adult male and female worms live in the abomasum, or true stomach, of ruminant animals. The female deposits 5000 to 10,000 shelled embryos per day, which pass out of the host in the feces. These undergo development on the ground, and the small first-stage juveniles hatch. The young worms feed on bacteria until they molt, shedding the cuticle and becoming second-stage juveniles. This stage also feeds on microorganisms for a time, and eventually molts the cuticle. However, this cuticle is not cast off but is retained as a sheath. These sheathed third stages do not feed and are infective for the vertebrate host. The ruminant becomes infected while grazing by eating the third-stage juveniles. Exsheathment occurs in the rumen, anterior to the abomasum, and the young worms pass into the abomasum where they burrow into the mucosa. Here they undergo another molt, and the fourth-stage juveniles come back into the paramucosal lumen of the abomasum. They begin to feed and undergo another molt before reaching adulthood. Mating of adults occurs and egg production commences. The life pattern is summarized in Fig. 6–1.

Development Outside the Host

The shelled embryos of *Haemonchus* are in the morula stage when they leave the host. Development to the first-stage juvenile requires that oxygen be available. The first stage worms emerge from the shell and begin to feed on bacteria. They grow during this feeding stage and then become quiescent (lethargus). The old cuticle is replaced from below by a new one, the old cuticle becoming detached from the body. This old cuticle is discarded as the worm again becomes active and begins to feed. The behavior of this second stage differs from that of the preceding one. The worm is more active and swims readily in water. As before, growth of the second-stage juvenile stops with the onset of a second lethargus, and the worm is quiet for a few hours. A new cuticle forms, and the old one separates from the epidermis. This old cuticle is retained as a sheath, and the juvenile nematode again becomes active, although it cannot feed. During this phase the juvenile worms apparently depend on stored body lipids as energy source and may live without feeding for as long as six months. This is the infective stage, and the worm is capable of surviving under conditions that would kill younger forms. Some years ago Glaser and Stoll cultured the free-living stages in a sterile medium containing liver extract, yeast extract, and agar. The sheathed juveniles were somewhat

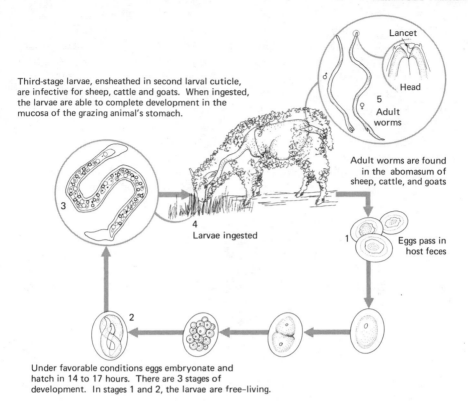

Third-stage larvae, ensheathed in second larval cuticle, are infective for sheep, cattle and goats. When ingested, the larvae are able to complete development in the mucosa of the grazing animal's stomach.

Lancet

Head

5
Adult
worms

Adult worms are found in the abomasum of sheep, cattle, and goats

3

4
Larvae ingested

1

Eggs pass in host feces

2

Under favorable conditions eggs embryonate and hatch in 14 to 17 hours. There are 3 stages of development. In stages 1 and 2, the larvae are free-living.

Fig. 6—1. The life pattern of *Haemonchus contortus*.

smaller than normal but were infective for the vertebrate host and yielded normal adult worms.

The movement of the infective juvenile on the ground is important, since infection of the vertebrate depends on the juvenile worms being eaten along with a meal of grass. These stages are commonly found on blades of grass, and it was once thought that this was due to a negative geotaxis. The movements of the juveniles are modified by such factors as temperature, humidity, and light intensity, but movement with respect to gravity is random. The juveniles tend to accumulate where climatic change is minimized.

Infection of the Vertebrate

When the sheathed juvenile of *Haemonchus* is swallowed by the host, exsheathment occurs in the rumen. Clearly the sheep rumen furnishes a stimulus for exsheathment. For twenty years, search was made for the unknown exsheathing substance in rumen fluid. The fluid was dialyzed, frozen, dried, aerated, boiled, filtered, and centrifuged, most such treatments resulting in loss of the exsheathment-stimulating activity. Rogers postulated that exsheathment involved two mechanisms, a stimulation from the host and, in reaction, the secretion of an exsheathment factor by the parasite. This was an important idea since it allowed for the consideration of substances that might not act directly on the sheath but would serve as a signal for exsheathment. Indeed, the chief factor in the stimulus turned out to be dissolved carbon dioxide or undissociated carbonic acid (or both). The effectiveness of the signal is modified by Eh, pH, and the presence of dissolved salts. Bicarbonate ion is not an effective signal.

Relatively high concentrations of carbon dioxide are required for the exsheathment of *Haemonchus,* and the process is enhanced by reducing agents, such as sodium dithionite, cysteine, or ascorbic acid. The absence of oxygen does not seem to be important per se. Various salts have been shown to affect the exsheathment induced by carbon dioxide, and their activity appears to be related to their capacity to catalyze the reaction

$$CO_2 + H_2O \rightleftarrows H_2CO_3.$$

The pH may have an effect that is independent of its effect on carbon dioxide solubility. Temperature of course has an effect on exsheathment. Increased temperature decreases carbon dioxide solubility but may enhance the reaction of the sheathed nematode to the stimulus.

If sheathed nematodes are exposed to carbon dioxide for a short time, say 15 minutes, and then washed and incubated for two hours in saline without carbon dioxide, exsheathment will proceed. If this saline is now collected and tested, it is found to have an effect on nematode sheaths quite independent of the carbon dioxide stimulus. This "exsheathment fluid" produced by carbon dioxide-stimulated juveniles contains a leucine aminopeptidase that acts on the sheath in an area about 20 μ from the anterior end. The anterior end then comes off like a cap. Interestingly, the composition of the sheath is such that it is not acted on by the proteolytic digestive enzymes of host origin nor by a leucine aminopeptidase from *Trichostrongylus,* a closely related nematode. The exsheathment enzyme from the juveniles requires Mg or Mn and is inhibited by sulfhydryl-binding agents.

The actual site of action of the host stimulus is not known. It might act through the nervous system or directly on the effector from which the exsheathment enzyme is derived. There is evidence that the enzyme is released at application of the stimulus, rather than being synthesized under the influence of the stimulus. It is probable that secretion occurs through what is anatomically termed the *excretory pore.*

The conditions required for exsheathment in vitro seem to be quite consistent with conditions in the rumen. The high carbon dioxide tension required for *Haemonchus* exsheathment would not commonly be found outside the rumen, and the pH, redox potential, temperature, and salt concentrations are appropriate. In the host, *Haemonchus* moves into the abomasum after exsheathment in the rumen. As noted above, it buries itself in the wall of the abomasum, undergoes another molt, and in about 40 hours can be found in the paramucosal lumen of the abomasum. In 1940, Stoll showed that the exsheathed third-stage juveniles will develop into the fourth stage in nonnutrient media in sealed tubes, and more recently Sommerville showed that this happens quite rapidly under 50 percent carbon dioxide. Hence it would appear that carbon dioxide is important in the further development of *Haemonchus* in the host. The role of carbon dioxide as a signal for differentiation, in contrast to exsheathment, merits further investigation, since it also induces certain events of differentiation in some other invertebrates, such as *Hydra.*

The Feeding of Adult *Haemonchus*

The pathogenic effects of *Haemonchus* have been attributed to its blood feeding habit. An infected lamb may lose as much as 175 to 250 ml of blood per day. Whitlock and his colleagues have made careful studies of the anemia that accompanies haemonchosis. Using ^{59}Fe labeling and whole body counting of sheep, as well as hematocrit values (packed red cell volume per unit whole blood), they have shown that the blood loss is directly correlated with the egg production of the worms. This relationship indicates that the blood contains some factor essential for egg biosynthesis in the *Haemonchus* system, and it has been suggested that this factor is oxygen. This is based on the fact that Crofton showed that (1) unless sufficient oxygen is available, egg development will not occur; (2) the shelled embryos deposited by the worm are in the morula stage, so some development has occurred; and (3) significant erythrocyte loss does not occur until the parasites are mature enough to deposit shelled embryos. These facts furnish circumstantial evidence that the worms acquire oxygen from the blood, all the more since Rogers showed some years ago that the oxygen tension in the paramucosal lumen of the sheep gut is low, and at the oxygen tensions prevailing there the worms respire at a rate that is about 12 percent of the maximum respiratory rate. Unlike *Fasciola* (p. 65), the evidence that *Haemonchus* feeds on blood seems very convincing. Undoubtedly the worm receives nutritional benefit from other components of whole blood, and, as Whitlock has remarked, oxygen is probably not a single, limiting factor.

Energy Metabolism of *Haemonchus*

There seems to be no information on the metabolism of the adult stages of *Haemonchus,* but some work has been done on the larval forms. A number of glycolytic enzymes are present in third-stage infective larvae, at levels resembling those of rat liver. An exception to this are very low levels of pyruvate kinase and lactic dehydrogenase. It has been suggested that the carboxylation of phosphoenol pyruvate is an important reaction in this parasite, leading from the glycolytic sequence to the Krebs cycle, but this has not been demonstrated. Most of the enzymes of the Krebs cycle have been found in the worm, but additional study is needed to ascertain the significance of the cycle as a series of energy-yielding reactions. The end products of carbohydrate metabolism are not known.

The Self-Cure Phenomenon

The self-cure mechanism has been demonstrated in several nematode parasitisms but seems to be best known in *Haemonchus.* It should be remarked that not all worm populations appear to be able to elicit the reactions associated with self-cure. Although such differences probably reflect significant differences in the protein makeup (antigenicity) of *Haemonchus,* the genetic constitution of sheep in different parts of the world might also be involved.

The self-cure reaction occurs when a dose of infective juveniles is given to a sheep already parasitized by *Haemonchus.* This results in a violent intestinal response in the sheep, which results in the loss of most or all of the population of adult worms in the host. In some cases the dose of larvae producing the effect may grow to maturity and set up a new worm burden, but in other cases all the worms, young and old, are swept out of the host. The final effect seems to be mechanical in nature. The edema and diarrhea produced in the self-cure reaction literally wash the parasites out of the host. The reaction occurs in sheep that have been exposed to fairly frequently spaced doses of infective juveniles; continuous infection with mature worms does not induce the necessary physiological conditioning. The reaction is triggered only by the ingestion of living juvenile worms; the administration of dead worms does not produce the response. The nature of the self-cure phenomenon is not adequately understood. It appears to resemble an allergic response. It is not set off instantly after administration of larvae, but appears at about the time that the juveniles are undergoing a molt in the wall of the abomasum. It has been suggested that the exsheathment fluid produced by the worm is the chemical stimulus in animals that have become sensitized to this antigen.

The Kinds of *Haemonchus*

As in the case of the *brucei* subgroup of trypanosomes, the biology of the worm that was long referred to as *Haemonchus contortus* turns out to be more complicated than at first appeared to be the case. Examination of some of the literature of a few years ago dealing with haemonchosis leaves the impression that people working on this parasitism in different parts of the world were actually working with separate parasitisms. Indeed they probably were and still are. For example, a superficial examination of the sheep in Britain, Australia, or upstate New York reveals that these animals are all quite distinct phenotypes. The selection of sheep for commercial purposes in these various localities has been very diversified and is the result of both climate and custom.

Studies on the genetic components of insusceptibility to *Haemonchus* have shown that sheep showing quite minor phenotypic differences, as measured by the superficial appearance of the animals, may have quite significant variations in apparent susceptibility. Further, there is now good evidence that there are quite definite contrasts between populations of *Haemonchus* in different sections of the world. The varied reactions of sheep to haemonchosis strongly indicate that there are antigenic distinctions between individual populations, and in some cases, certain populations of *Haemonchus* fail to elicit the self-cure phenomenon.

Eventually, morphological differences between particular nematode populations were discovered, and subspecific names have been given to some of these forms. Other significant biological deviations have also emerged. Subspecies of *Haemonchus contortus* show modifications in generation time, including the time needed to reach maximum egg production in the vertebrate host, minimal egg hatching time, and the cultural requirements of juvenile stages. In an analysis of the literature, Crofton and Whitlock were able to adduce evidence that the segregation of such characteristics, resulting in shifts of phenotype boundary, could have occurred in as short a period as 40 years.

A Systems Approach to Haemonchosis

Ratcliffe and his colleagues have applied a systems-analysis approach to the *Haemonchus*-sheep system. As already indicated, the major lesion in haemonchosis is an anemia that can be assessed by hematocrit determi-

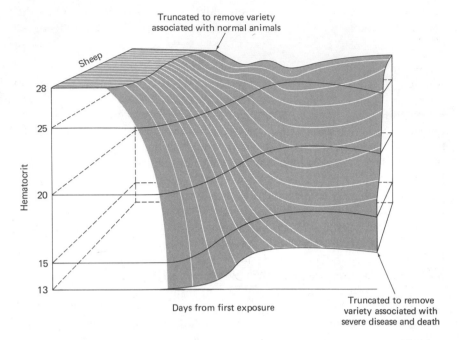

Fig. 6–2. Generalized model of the variations in the hematocrit trajectories of different sheep infected with *Haemonchus contortus*. Hematocrit is the volume of packed red cells per unit volume of blood and is thus a measure of anemia. (After Ratcliffe et al., 1969.)

nations. There is some variation in the extent of disease suffered by individual animals (Fig. 6–2). It is also known that there is an inverse relationship between hematocrit value and egg production of the parasite. Although the variation among sheep presents some problems, for purposes of analysis the problem may be generalized.

It is necessary to postulate threshold values for individual animals in a model of the mechanisms involved in the parasitism (Fig. 6–3). This model takes into account the factors known to be involved in the life cycle of the nematode but also recognizes that there are gaps in the specific information available on the parasitism. The continuous lines shown in Fig. 6–3 represent flows of real quantities, whereas the dashed lines represent flows of information. Variables that must be specified in the actual input for a computer program are enclosed in dashed boxes. The S-shaped curves indicate instances in which a provision has been made for a function relating two variables. When actually programmed, these S-shaped relationships are simulated by a single subroutine that can be used in computing a great variety of potentially S-shaped functions. The exact form of the function must be specified in each case. It is important to note that in

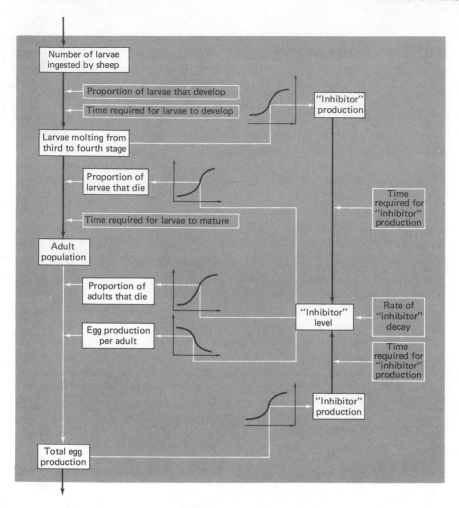

Fig. 6—3. A postulated model of the parasite control mechanisms in haemonchosis. See text for discussion. (After Ratcliffe et al., 1969.)

a particular relationship the variable must exceed a certain trigger value, or the relationship becomes void. The sum of influences is denoted by the single abstract "inhibitor," which is satisfactory for analysis.

A suggested model for the regulation of the hematocrit in haemonchosis is presented in Fig. 6–4. The symbols used are similar to those used in Fig. 6–3. In the actual computation the "total erythrocyte volume" is set at an initial value. Thereafter it is updated according to the amount of erythrocyte loss (remembering the proportionality to egg production by the worms) or the amount of erythrocyte production. The hematocrit level is computed from an empirical equation derived from his-

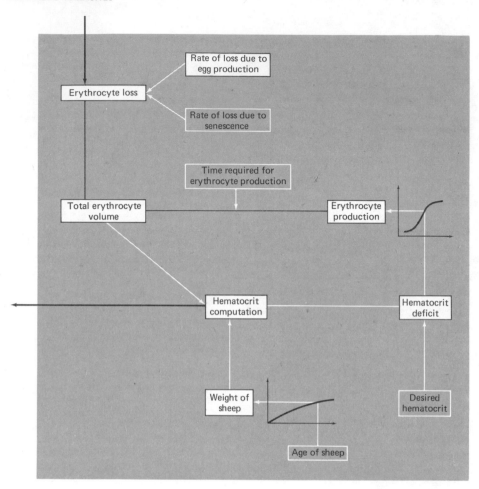

Fig. 6–4. A postulated model of hematocrit regulation in haemonchosis. See text for discussion. (After Ratcliffe et al., 1969.)

torical data. The subroutine for generating S-shaped curves is used at this point for describing weight as a function of time. Although there may be some feedback relationship between the weight of the sheep and other components in the system, no allowance was made for this in the model.

Comparison of this model with live sheep allowed a realistic explanation of the life-death expectancy of an individual sheep based on (1) the maximum capacity of an individual sheep to produce erythrocytes, and (2) the relationship between erythrocyte production and erythrocyte deficiency in an individual sheep. The authors showed that a simulation model can be used as a basis for selecting the specific variables that should be studied and for estimating these variables. The above discussion is highly abridged, and the original paper should be consulted.

TRICHINELLA SPIRALIS

The trichina worm has continued to receive man's attention for more than a hundred years. The sometimes fatal result of human trichinosis in the temperate and cold regions of the world has insured this attention. In the United States, it is now considered an exotic disease, although we may anticipate outbreaks as long as individuals slaughter their own pigs or eat undercooked bear meat. Since most trichinosis in humans has been attributed to the eating of poorly cooked pork, it has sometimes been said that the Hebrew prohibition of pork is an example of Mosaic wisdom applied in controlling an infectious disease. However, the prohibition is probably due more to the difficulty of keeping pork in an edible condition in hot climates without refrigeration, and to the solemnizing of pigs as totem animals. As will become apparent, man is not a "natural" host for the worm, and in reality he represents a dead end in the cycle.

The Life Pattern of Trichinella

The juveniles of *Trichinella* are eaten as encysted forms in meat. They are liberated from their cysts by the action of the digestive enzymes of the host and penetrate a short distance into the mucosa of the small intestine. Here they molt and reach adulthood in 40 to 50 hours. Mating occurs, and weakly shelled vermiform embryos develop in the uterus of the female. The juvenile worms are deposited in the mucosa and, within a week after initial infection, may be found in the lymph or blood of the host. They undergo further development after entering skeletal muscle fibers, where they become encysted. Here they grow rapidly, molting and attaining the infective juvenile form in about three weeks. If this infected muscle is now eaten by a vertebrate, the cycle is repeated. The life pattern is summarizd in Fig. 6–5.

The Intestinal Phase of Trichinosis

During the early phase of *Trichinella* infection, the worms burrow into the mucosa of the host. It might be thought that they would possess some enzymatic factor for entering, such as has been found to be the case in worms like *Schistosoma* (p. 75). However, all attempts to demonstrate such chemical aids to penetration have been negative. Apparently the worms do not pass through the basement membrane of the mucosa, although the mechanics of their burrowing is not yet understood.

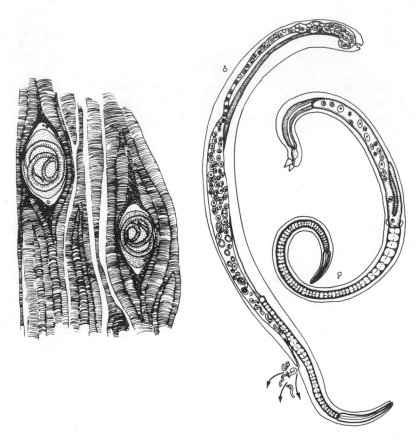

Fig. 6—5. The adults (left) and encysted muscle larvae (right) of *Trichinella spiralis*. See text for discussion of the life pattern.

During this phase the host shows rather marked symptoms of an inflammatory character in the gut. Diarrhea and pain are often associated with the infection. There is good evidence that the presence of the worms interferes in the digestion of protein and with the absorption of sugars and calcium from the digestive tract. It has been suggested that these interferences with digestive tract function are rather general in nature and may be quite similar in origin to disturbances seen in hookworm disease, nontropical sprue, and niacin deficiency in dogs (Fig. 6–6). Perhaps they are diseases of the brush border of the vertebrate intestine. There is modification of the intrinsic brush-border hydrolases, as well as of the membrane transport systems involved in the absorption of low molecular weight organic compounds. Further, there is morphological alteration of the intestinal villi (Fig. 6–7). A meticulous study of the effects on ion fluxes would be desirable, particularly in view of the relationship that is

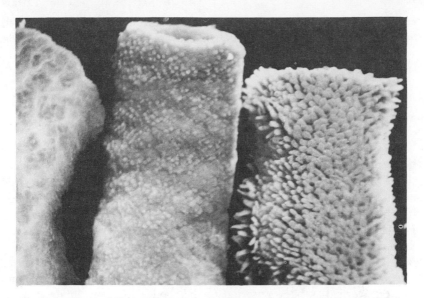

Fig. 6–6. Effects of the intestinal phase of *Trichinella spiralis* on the small intestine of the mouse. Each intestine is everted. (Right) Normal mouse. (Middle) Hobnail atrophy seen on the fourth day after infection. (Left) Convoluted-mosaic atrophy seen on the tenth day after infection ($\times$ 18). (Courtesy of Dr. Leroy Olson and the *Journal of Parasitology.*)

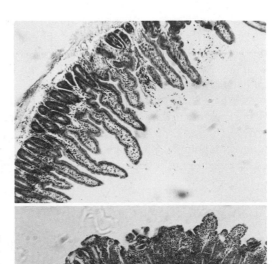

Fig. 6–7. Effects of trichinosis on the villi of the anterior small intestine of the mouse. Cross sections of upper small intestine. (Upper) Villi from an uninfected mouse. (Lower) Atrophied villi from an infected mouse ($\times$ 103). (Courtesy of Dr. Leroy Olson and the *Journal of Parasitology.*)

recognized between glucose absorption and the entry of sodium in the small intestine. The observed intestinal symptoms do not last very long in trichinosis. In experimental animals, intestinal effects disappear in a few weeks, since the life of the adult worms is quite abbreviated.

Metabolism of *Trichinella*

Nothing is known of the metabolism of the adult forms of this worm. However, the metabolism of juvenile forms recovered from mammalian muscle has been studied. The juvenile worm is well supplied with glycogen stored in its tissues, and in vitro, it utilizes this glycogen. Under either aerobic or anaerobic conditions, glycogen is fermented to a mixture of volatile fatty acids. The most prominent of these is *n*-valeric acid, which constitutes more than 50 percent of the acid produced. In addition, lesser quantities of C_2, C_3, C_4, and C_6 acids are formed and excreted. Very small quantities of lactic acid are produced. Under aerobic conditions the worm has been reported to metabolize its body lipids, but this has not been studied in any detail. The general impression is one of an essentially anaerobic metabolism. Certainly the products of energy metabolism are only partially oxidized, as seems to be the case with the quite unrelated forms that were discussed in preceding chapters.

Six of the enzymes of the Krebs cycle have been identified in juvenile *Trichinella*. Further, the worms appear to have some form of cytochrome system. Assays for cytochrome components have shown them to be present, and the respiration is inhibited by cyanide, which is consistent with this view. However, respiration is not inhibited by carbon monoxide, and this is somewhat puzzling.

When *Trichinella* juveniles were incubated in vitro, Haskins and Weinstein found that they excreted ammonia, several amino acids, and a series of amines into the medium. The amines included at least nine different compounds, some of which have pharmacological activity and might be involved in the production of symptoms in trichinosis.

Changes in the Infected Muscle

During the early stages of invasion of skeletal muscle by the juvenile worms, there is an increase in muscle glycogen. This is followed by a sharp decrease in glycogen. Profound changes in the energy metabolism of the muscle clearly occur. There is a decrease in respiration, and there appear to be modifications of the Krebs cycle; increased fat synthesis is observed. Infected muscle in hogs has been shown to contain *n*-valeric acid, which is not a normal metabolite of vertebrate muscle but is known to be produced by *Trichinella*. The incorporation of radioactive inorganic phosphate into high-energy phosphate compounds, such as ATP, increases, indicating an increased rate of phosphate turnover.

There are modifications of amino acid metabolism in the infected muscle. The ^{14}C of radioactive tyrosine is incorporated into protein in infected muscle at a much lower rate than into protein in uninfected muscle. On the other hand, ^{14}C of radioactive tryptophan is incorporated into protein in infected muscle at a rate ten times more than that in which it is incorporated into protein in uninfected muscle. These changes are not well understood, but they indicate a sharp reorientation of metabolism.

The *Trichinella*-Host Cell System

Much of the older literature dealing with trichinosis refers to the interaction of *Trichinella* with the host muscle fiber as "degenerative," referring primarily to what was thought to occur in the fiber. However, this concept has required revision, with the recognition that the changes that occur in the infected fiber can be considered to be dedifferentiation and redifferentiation with the loss of specialization for contractile function and with the appearance of the new characteristics of a host cell-parasite complex.

It has been shown that shortly after the juvenile worm enters the muscle fiber, there is a burst of RNA synthesis in the parasite, followed by a burst of RNA synthesis in the host cell. The myofibrils undergo disorganization and disappear from the muscle fiber. This is accompanied by an increase in the rough endoplasmic reticulum and in mitochondria in the host cell. The worm undergoes growth. Lyosomes appear in the fiber during this period, and various lysosomal enzymes are readily identified in the fiber at this time. The Golgi complex is enlarged, and the metabolism of the host cell is now clearly directed to the synthesis of the collagen-glycoprotein cyst around the larva and to the support of the parasite and the modified fiber. It is now of course no longer a muscle fiber and represents a completely different entity. It does not "degenerate" in hosts in which the worm undergoes development.

There is one host, however, in which the muscle fiber indeed does degenerate. Ritterson was surprised to find that after infecting the Chinese hamster and obtaining quite good development of the intestinal phase, very few larvae developed in the muscle. In this host the presence of the worm does cause a rapid degeneration of the infected muscle fiber. No reconstitution as described above occurs, and the larval worms cannot survive. This degeneration in the Chinese hamster can be arrested by the administration of cortisone, and in such treated animals there is quite normal development of the muscle phase of *Trichinella*.

After a few months, the cyst surrounding the juvenile *Trichinella* begins to undergo calcification (Fig. 6–8). As a matter of fact, the medical student who first discovered *Trichinella* in a dissecting room in London noticed the granular nature of the calcified cysts in a heavily infected human cadaver. The cyst itself may become quite heavily calcified before the juvenile worm dies and becomes calcified. The nature of the calcification process is not known. However, the administration of parathormone speeds up calcification of *Trichinella* cysts and of the coronary arteries of the host in parallel fashion, so it may be assumed that calcification occurs by mechanisms similar to those involved in other calcification processes of the host. (This also means that speeding up calcification by artificial means seems to hold little promise as a way of treating trichinosis.)

The Kinds of *Trichinella*

During the past few years several workers have reported that *Trichinella* from different parts of the world showed variable capacities to infect laboratory animals. In the late 1950s the present author and Dr. Everett Schiller discovered that a *Trichinella* isolated from polar bears in the American Arctic failed to develop in laboratory rats, although these rodents were easily infected with *Trichinella* isolated from hogs on the Gulf Coast of the United States. A few years later British workers found that a strain of *Trichinella* from Kenya showed very low infectivity for rats or domestic pigs and verified the findings of differences between forms from Alaska and

Fig. 6–8. Encysted larvae of *Trichinella* in striated muscle of a pig.

those occurring in Europe. It now seems clear that there are real strain distinctions in *Trichinella* from various localities, and the biology of the organisms may differ in other ways. It would be of great interest to study the forms that are found in the walrus, since there is presently no satisfactory way of explaining how walruses become infected.

Human Trichinosis

The general geographical pattern of human trichinosis is correlated with the patterns of human nutrition. Man tends to be a predominantly herbivorous animal near the equator; in the temperate zones he has been and is an omnivorous animal. In the polar regions man is a carnivore, except for those modern instances in which he can import vegetable food from areas nearer the equator. Since the transmission of trichinosis is completely dependent on the carnivorous habits of hosts and since the parasite seems to infect a great variety of mammals, it may be expected that parasite incidence in a given mammalian host would be correlated with the tendency to eat other mammals. Thus, a few years ago there was an outbreak of human trichinosis in Washington, Missouri, associated with meat-eating and involving 30 or more people, and at least one Arctic expedition has failed because the explorers ate inadequately cooked bear meat and died from the effects of trichinosis. The incidence of human trichinosis, therefore, is higher in the temperate zones and in the polar regions than in the tropics. In the United States the general incidence of infection was estimated a few years ago to be around 16 percent, although it may be lower now.

From the number of juvenile worms found at autopsy in individuals who had had no reported clinical history of the infection, it is clear that the disease must have been often overlooked or mistaken for something else. Cases have been variously diagnosed as typhoid, dysentery, amoebiasis, viral gastroenteritis, food allergy, and a number of other things. Also, in a great many instances, the infection may have gone unnoticed by a patient who "just didn't feel well." The seriousness of the symptoms associated with the infection is a function of the size of the dose of juveniles ingested by the host, and some of the heavier muscle burdens discovered at autopsy may represent small doses repeatedly ingested over a period of time.

In acute trichinosis the host responses can be separated into three phases that correspond to the life pattern of the worm. The first phase is of course intestinal, and it may be accompanied by weakness and muscular twitching. As the juveniles begin to be deposited by the female worms, eosinophilic leucocytes appear in the host blood, sometimes comprising as many as 90 percent of the white cells.

The second phase corresponds to the period of juvenile worm migration and muscle penetration. There is a characteristic edema (marked puffiness) about the eyes, and there is intense muscle pain. As the muscles are invaded there may be disturbances in the function of particular muscles, causing difficulty in moving the eyes, respiration, and such. Respiratory difficulties become more marked in the fourth and fifth weeks. Fever is suffered more or less constantly during this second phase. The eosinophilia noted earlier persists.

The third stage is associated with the beginning of encystment of the worms in the muscles. The symptoms of the second phase become more marked. The face becomes puffy, and there is a tendency for swelling of the extremities. Pneumonia is commonly a complication in heavy infections. Damage to the heart, to the nervous system, and to other organs may cause additional symptoms. This damage is apparently due to earlier wandering of the juvenile worms, since they do not stay in these organs. If the host recovers, there may be recurring muscle weakness and pain for a year or more.

Humans become infected from eating raw or uncooked meat, most often pork. In the past, pigs became infected from eating garbage containing pork scraps, and it was shown some years ago that the frequency of pig infections is closely related to the methods of feeding pigs. Occasional outbreaks occur from eating other kinds of meat, such as bear meat. It is clear that all human infections could be prevented if meat were thoroughly cooked. However, domestic trichinosis can also be controlled if the garbage fed to hogs is cooked. Most states now require that the garbage destined for feeding hogs must be cooked, and there is no doubt that the incidence of human trichinosis in the United States has been declining. It took many years of effort to obtain the necessary legislation for this step. The hog raisers of the nation were highly resistant to this at first, but a serious and expensive outbreak of vesicular exanthema, a pig disease transmitted through garbage feeding, finally persuaded the swine industry that it would be in their interest to support legislation requiring the cooking of garbage that was fed to hogs.

ASCARIS

The large nematode parasite of man and pigs, *Ascaris lumbricoides,* may very well have the dubious distinction of being the first human parasite recognized as such by primitive man. It is of a size that makes it rather obvious, the female having a length of 8 to 12 in. with a diameter of 4 to 6 mm. The males are somewhat smaller (Fig. 6–9). It has long been a favored animal for the study of nematode anatomy by undergrad-

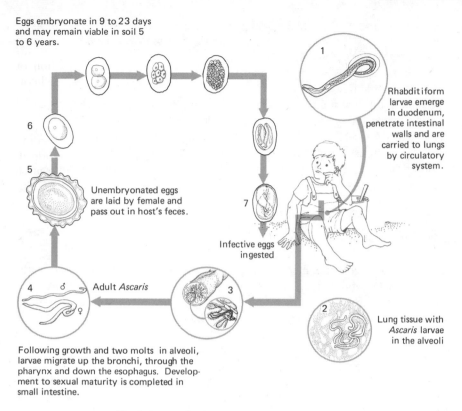

Eggs embryonate in 9 to 23 days and may remain viable in soil 5 to 6 years.

1

Rhabditiform larvae emerge in duodenum, penetrate intestinal walls and are carried to lungs by circulatory system.

6

5 Unembryonated eggs are laid by female and pass out in host's feces.

7

Infective eggs ingested

4 ♂ Adult *Ascaris* ♀

3

2 Lung tissue with *Ascaris* larvae in the alveoli

Following growth and two molts in alveoli, larvae migrate up the bronchi, through the pharynx and down the esophagus. Development to sexual maturity is completed in small intestine.

Fig. 6—9. The life pattern of *Ascaris lumbricoides.*

uates because of its large size and ready availability, in spite of the fact that it is not a particularly "typical" nematode. The form in pigs differs physiologically from that in man, the larvae of worms from one host species not readily infecting the other.

The adult *Ascaris* most often lives in the small intestine of the vertebrate host and appears to feed on the contents of the gut lumen. The worm would appear to have a voracious appetite; it has been estimated that the weight of the 200,000 shelled embryos and accompanying secretions produced by a female *Ascaris* in 24 hours is equal to about 10 percent of the worm's body weight. It is perhaps not surprising that the worm interferes with protein nutrition in the host, and this is of considerable medical significance in the protein-poor tropics where the man-*Ascaris* parasitism is quite common. In 1964, it was estimated that 26 percent of the human population on earth was infected with *Ascaris.*

Historically, *Ascaris* has been of great importance in biology. As early as the seventeenth century, Francesco Redi used his studies on

Ascaris to attack the widely held belief in the spontaneous generation of organisms, and for years the egg and embryo of *Ascaris* have been classical materials for cytologists.

Some time ago the eminent biologist, Richard Goldschmidt, wrote a book entitled *Ascaris—the Biologist's Story of Life.** Goldschmidt's point was quite simply that most of the general concepts of biology could be demonstrated using this roundworm parasite as an example. Indeed, although the organization of *Ascaris* is deceptively simple, much research on this organism has been carried out since the appearance of Goldschmidt's book, and the worm has continued to furnish new surprises for research workers.

The Life Pattern of *Ascaris*

The shelled unsegmented embryos leave the mammalian host in the feces. After about three weeks on the ground, the juvenile worms have become vermiform, have molted, and are capable of infecting a vertebrate. If they are swallowed, the young worms hatch and penetrate the mucosa. They get into the circulatory system of the host and are carried to the lungs. Here they break out into the alveoli and are passively carried out of the lungs to the trachea and esophagus. After being swallowed, they remain in the small intestine and develop to adulthood in about two months. Adult life is about nine months. This pattern is shown in Fig. 6–9.

Embryonation

The unsegmented embryos of *Ascaris,* enclosed in a multilayered shell, are deposited by the female worm and pass out of the host in the feces. For development of the embryo, oxygen, water, and a temperature well below 98° F are required. About 85° F seems to be the most favorable temperature for development. The requirement for oxygen may seem paradoxical, since oxygen has a markedly toxic effect on the adult worms. This is suggestive of some very fundamental biochemical differences between the embryos and adults, which will be discussed presently.

Under favorable conditions, *Ascaris* embryos develop to a stage infective to the vertebrate host in about three weeks. In 1957, Passey and Fairbairn showed that a net synthesis of carbohydrate occurred during the development of the shelled embryo of *Ascaris* outside the host (see Fig. 6–10). The increase in the fat-free dry weight of embryos paralleled

* See references at end of this chapter.

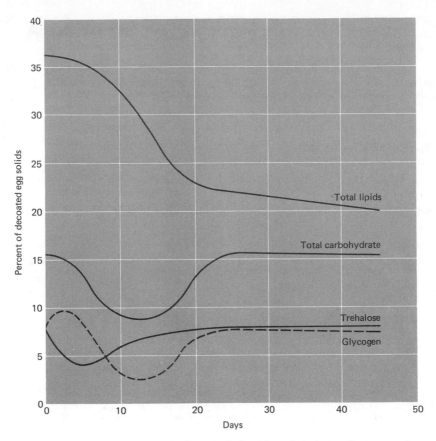

Fig. 6–10. Changes in total lipid, total alkali-stable carbohydrate, glycogen, and trehalose during development of shelled *Ascaris* embryos at 30° C. (After Passey and Fairbairn, 1957.)

the increase in glycogen and the disaccharide, trehalose. The amount of lipid carbon disappearing during sugar synthesis was equal to the amount of carbon accounted for as carbon dioxide plus sugar synthesized. This appears to be a clear demonstration of the synthesis of carbohydrate from fats, a phenomenon long hypothesized but not previously proven to occur in animal cells. During this period of development the shell is astonishingly impermeable, not allowing the passage of salts, although water and dissolved gases pass through the shell. If, however, the mature embryo is ingested by a mammal, a rapid change takes place. A combination of high carbon dioxide tension and low oxidation-reduction potential operating at a pH around neutrality at 98° F triggers a physiological change in the embryo. A rapid alteration of permeability occurs, followed by dissolution of the outer layers of the shell. This dissolving of shell material is attributable to secretions from the young worm itself. An esterase and a chitinase have been identified in the hatching secretion of the worm. The effect of carbon dioxide in inducing hatching of the shelled embryos of intestinal worms is a common phenomenon. Seven species of ascarid worms and four species belonging to other families have been shown to hatch with carbon dioxide stimulation.

Development in a Vertebrate

After emerging from the shell the young *Ascaris* penetrates the mucosa of the host. Although the physical and/or chemical mechanisms are unknown, penetration must occur, since the worms pass on into the vascular system of the host. In the blood system the immature worms are carried, probably passively, first to the heart and then through the pulmonary arteries to the lungs. During this migratory phase the worms increase in length about tenfold from an initial length of about 250 μ, but little is known of the physiology and biochemistry of this stage of parasite development.

In the lungs the young *Ascaris* escape from the capillary bed into the alveoli and are carried in the mucus stream to the bronchi, trachea, and eventually to the pharynx. If swallowed by the host, the worms continue development in the small intestine, requiring about two months to attain sexual maturity; sexual reproduction ensues. It might be expected that this migration through the lungs of the host would constitute considerable trauma. However, the host normally would ingest only a few embryonated eggs at a time, and the number of worms in the lungs at any given time would be small. (A notable exception to this was observed in the 1970 case of a Canadian graduate student who fed his roommates a very large single dose of embryonated *Ascaris* eggs. In this instance the dosed individuals almost died.)

Ascaris will hatch and carry out the migration just described, at least to the lungs, in a variety of mammals, including rats, mice, and guinea pigs, but will not develop to adulthood in such hosts. Only in humans or pigs is development completed.

The significance of the peculiar migratory pattern in the life history of *Ascaris* remained mysterious until it was shown that many ascarid parasites of carnivorous hosts migrate into the extraintestinal tissues of rodents and other small mammals, where they remain in a state of suspended development until the small mammal is devoured as prey. These ascarids then develop to sexual maturity in the carnivore. It may be suggested that in the evolution of *Ascaris lumbricoides,* critical events of development are genetically linked to the characteristics of behavior resulting in the migratory pattern.

Nutrition of the Adult

The evidence available suggests that the adult worm feeds on the contents of the intestinal lumen. The structure of the outer cuticular surface indicates that this is not differentiated as an absorptive surface, and indeed it has been shown that the cuticle is essentially impermeable to some ions and to sugars and amino acids. Apparently the gut serves digestive as well as absorptive functions. A number of hydrolytic enzymes have been identified in extracts of *Ascaris* gut, although it is not known whether all of these are secreted into the gut of *Ascaris*. The luminal gut surface in *Ascaris* is covered with microvilli, suggesting its absorptive function (Fig. 6–11), and it has been shown that the gut cells transport glucose against a cencentration difference. Interestingly, the anthelminthic drug dithiazanine blocks glucose transport by *Ascaris* gut, but it does not significantly affect sugar transport by host mucosal cells. This difference in membrane transport systems has not been further investigated. There is also evidence that amino acids and fatty acids are absorbed by

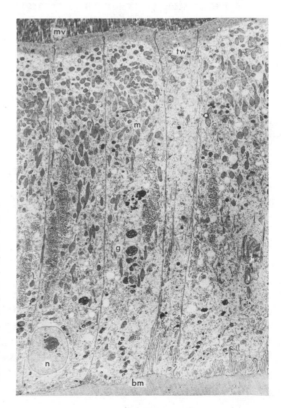

Fig. 6–11. The intestine of *Ascaris*. Note that this tissue is a monolayer of cells, making it peculiarly advantageous for examining cellular transport. The upper layer of microvilli (mv) faces the intestinal lumen, and the hemocoele fluid is immediately adjacent to the basement membrane (bm). Other abbreviations: nucleus (n), mitochondria (m), glycogen (g), terminal web (tw) ($\times$ 1300). (Furnished through the courtesy of Dr. Harley Sheffield and the *Journal of Parasitology*.)

Ascaris gut cells by specific transport mechanisms.   Hence the gut appears to be the major organ for digestion and absorption of food.   As will be noted below, it may also have an excretory function.

Carbohydrate Components of *Ascaris*

A considerable amount of information has accumulated on the carbohydrate metabolism of *Ascaris,* and in this context some very interesting observations are available on the carbohydrate components of the tissues of this parasite.   Of the sugars available in the hemolymph, the major one is the disaccharide trehalose.   Trehalose is synthesized from glucose by the reproductive organs and by the muscle of *Ascaris,* but not by the intestine, which contains a trehalase.   Glucose is present at very low concentrations.   N-acetylglucosamine and N-acetylgalactosamine are constituents of glycoproteins in *Ascaris*.   Glycogen is a major component of the tissues, and when extracted by mild cold-water procedures, it is found to be polydisperse, with the two main elements having molecular weights of about 5×10^7 and 4.5×10^8, respectively.

An interesting sugar known as ascarylose (3,6 dideoxymannose) has been isolated from *Ascaris*.   This is of particular significance because the dideoxyhexoses have generally been regarded as characteristic components of bacteria.   The sugar is present in what was formerly termed "ascaryl alcohol," but this is in reality a mixture of three glycosides having the configurations shown in Fig. 6–12.

Fermentation Products of *Ascaris*

The adults of *Ascaris lumbricoides* produce a witches' brew of fatty acids as end products of energy metabolism.   Included are α-methylvaleric, *n*-valeric, cis-α-methylcrotonic (tiglic), propionic, lactic, butyric, and C_6 acids, probably including *n*-hexanoic.   The worms may also produce acetylmethylcarbinol and ethyl alcohol.   *Ascaris* appears to use enzymes of the classical Embden-Meyerhof sequence in the degradation of glucose to the level of phosphoenol pyruvate.   However, classical glycolysis seems to be a minor portion of metabolism, since lactic acid is produced in very small amounts.   Pyruvic kinase is very low in *Ascaris,* and lactate may be formed by decarboxylation of malic acid, followed by the reduction of pyruvate.

The tricarboxylic acid is not a significant operating component of energy metabolism in adult *Ascaris*.   However, some intermediate compounds, ordinarily identified with the tricarboxylic acid cycle, play important roles in energy metabolism.   Succinic acid is produced in *Ascaris*

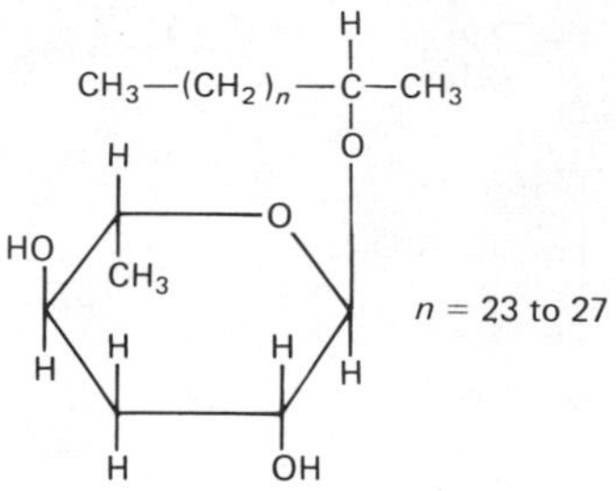

Ascaroside A

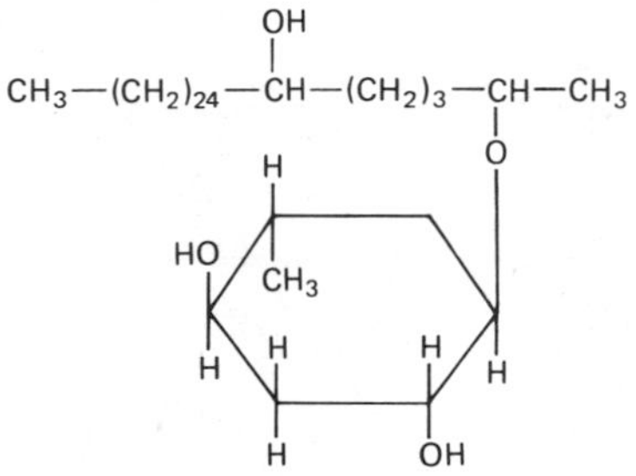

Ascaroside B

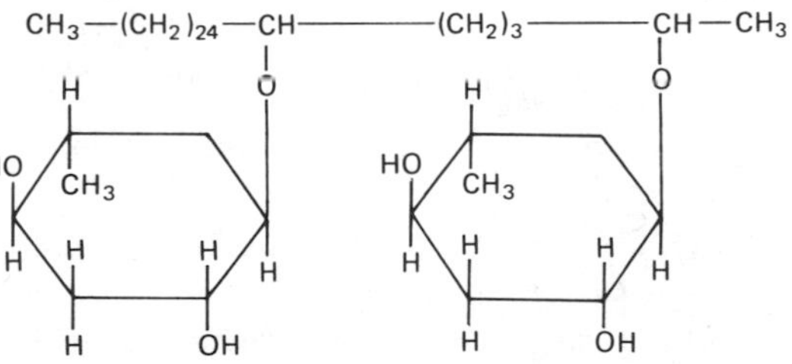

Ascaroside C

Fig. 6–12. Configuration of ascarosides A, B, and C from the tissues of *Ascaris*.

tissues.   There is some evidence that this occurs by carbon dioxide fixation, resulting in the formation of oxaloacetic or malic acids with subsequent reduction to succinic acid.   Succinate is produced during anaerobic incubation of an *Ascaris* muscle preparation with NADH and fumarate, and the system appears to be localized in the mitochondria.

Propionic acid appears to be produced by succinate decarboxylation, and propionyl-Coenzyme A probably figures prominently in the production of some of the exotic fatty acids produced by *Ascaris* (Fig. 6–13).

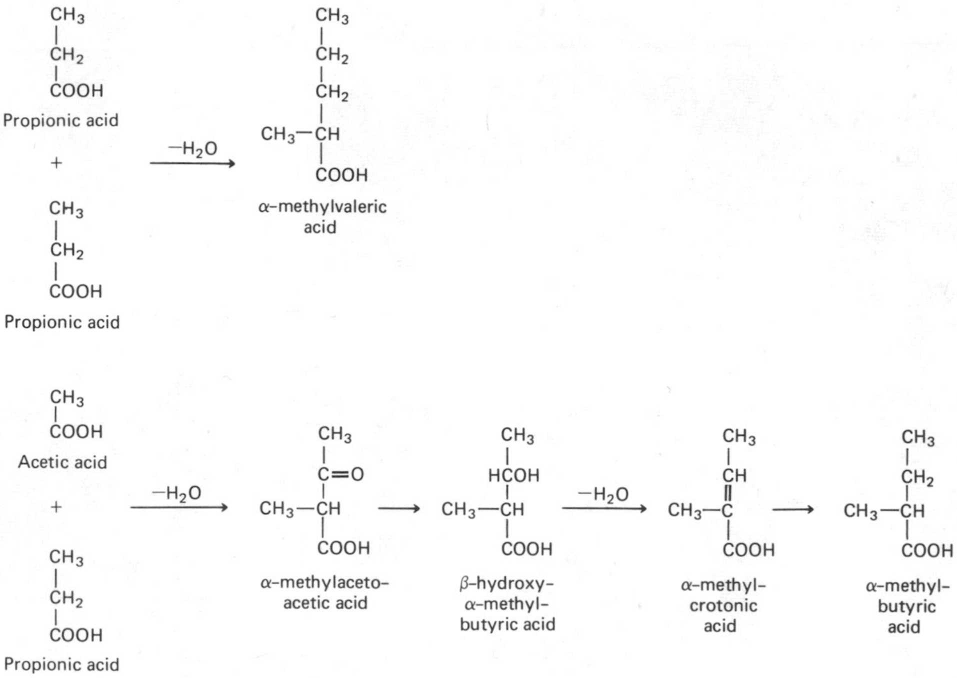

Fig. 6–13. Suggested mechanisms for the formation in *Ascaris* of α-methylvaleric acid and α-methylbutyric acid.

The mechanism for carbon dioxide fixation in *Ascaris* has not been adequately clarified. The organism has a malic enzyme, but there is evidence that phosphoenolpyruvic carboxykinase is of considerable importance. The production of acetylmethylcarbinol may involve the decarboxylation of pyruvic acid to form an acetaldehyde intermediate, which then combines with another molecule of pyruvic acid to form acetylmethylcarbinol.

The major excretion of fatty acids apparently occurs by way of the digestive tract. This simple cellular system (Fig. 6–11) seems to have multiple functions that may be very significant in determining the evolutionary success of nematodes in exploiting a great variety of habitats with minimal morphological change.

Lipid-Carbohydrate Conversions

As noted on p. 130, there is an apparent net aerobic conversion of fat to carbohydrate in the developing shelled embryo of *Ascaris*. It has been demonstrated that CO_2 is not extensively fixed in the process, and it was suggested by Fairbairn and Passey that partially oxidized fatty acids were involved in the conversion. Subsequently it was shown that the following volatile fatty acids were released by saponification of neutral lipids of the early embryo—α-methylbutyric, α-methylvaleric, and acetic acids—with propionic, formic, butyric, and valeric acids also released to a lesser extent. These are, as indicated, fermentation products of adult tissues. As embryonic development proceeds, the α-methylbutyric and α-methylvaleric acids of the neutral lipid decrease markedly. The energy source relationships of these conversions were rationalized, in part, by Saz and Lescure as shown in Fig. 6–14. It should be pointed out that volatile fatty acids comprise less than 20 percent of the lipid fatty acids used in the lipid to carbohydrate conversion. At least half of the fatty acids are nonvolatile ones, and recent studies indicate that these are probably obtained from intestinal contents of the host.

Electron Transport and Morphogenesis

There is evidence that the Krebs cycle is operable in the embryos of *Ascaris* during development outside a host. On the other hand, the Krebs cycle seems to operate at a very low level in the tissues of the adult worm, and it is probably of no energetic significance. Thus the major end products of energy metabolism in the developing worm are carbon dioxide and water, whereas the adult produces reduced carbon compounds

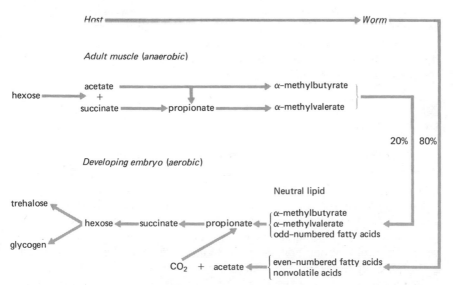

Fig. 6–14. Suggested relationships between fatty acids derived from the host and from synthesis in adult worm muscle and the synthesis of carbohydrate in the developing embryo of *Ascaris*. (After Saz and Lescure, 1966, in part.)

as end products. These patterns are closely related to differences in the predominating pathways for electron transport in the young developing worm and in the adult.

Although it contains some cytochrome components at low concentration, the adult *Ascaris* does not seem to have a functional cytochrome system, and it is either an anaerobic or microaerophilic organism. According to Bueding, the major pathways for electron transport involve flavoproteins and terminal flavin oxidases. In the undeveloped embryo, cytochrome *c* is present at very low concentrations and is in the reduced state. As development proceeds to the gastrula stage, cytochrome oxidase appears at low levels, and respiration becomes sensitive to antimycin A, an inhibitor that blocks electron transport at the cytochrome *b* stage in other tissues. Cytochrome *c* is present in a partially oxidized state at this stage, and the reduced fraction is increased by the addition of cyanide or carbon monoxide. During the vermiform stages, carbon monoxide continues to produce increased cytochrome *c* reduction and to inhibit respiration, and the effects of carbon monoxide are reversed by light. All of these data are consistent with the view that a cytochrome system develops in the embryonic stages of *Ascaris*. The presumed pathways in adults and developing larvae are illustrated in Fig. 6–15.

It has been shown that larval *Ascaris,* recovered from the lungs of guinea pigs eight days after infection, have an aerobic metabolism. Unlike the adult worm, these larval forms show a Pasteur effect, they are inhibited by cyanide or anaerobiasis, and they possess cytochrome oxidase, suggesting that the change from metabolism of the larval type to that of the adult type occurs after the worm has been in the host for several days. More recently it was demonstrated that the adult pattern of metabolism is apparent after the fourth molt; this corresponds to the time when the worms move from the lungs to the intestine for development as adults.

Nitrogen Metabolism

The rate of synthesis of new tissue carried on in a female *Ascaris* (see p. 129) would suggest high rates of protein synthesis, but we know surprisingly little about it. Alanine and aspartic acid can be synthesized by reductive animation of pyruvate or oxalacetate. Reduced NAD is required in the reactions. A number of transaminations have been described, with various amino acids including alanine and aspartic acid serving as amino group donors. As already indicated, the worm also possesses specific transport mechanisms in the gut for the absorption of amino acids.

There is no evidence that adults or developing embryos of *Ascaris* utilize proteins as an energy source. However, the worms produce a

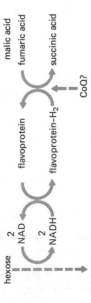

hexose

2 NAD / 2 NADH flavoprotein / flavoprotein-H_2

malic acid fumaric acid succinic acid

CoQ?

The anaerobic pattern of electron transport in adult *Ascaris*. Coenzyme Q may also participate in this system.

flavoprotein / flavoprotein-H_2 CoQ / CoQ·H_2 2 Cyto b / 2 Cyto b·H 2 Cyto c_1 / 2 Cyto c_1·H 2 Cyto c / 2 Cyto c·H 2 Cyto a / 2 Cyto a·H 2 Cyto a_3 / 2 Cyto a_3·H H_2O

$2H^+$

The aerobic pattern of electron transport in larval *Ascaris*.

Fig. 6–15. Electron transport patterns postulated to predominate in adults and developing larvae of *Ascaris*.

variety of products of amino acid metabolism. Most of the nitrogen excreted is in the form of ammonia. However, urea is also excreted, and urea synthesis is stimulated by the addition of arginine, citrulline, and ornithine, suggesting that some form of a Krebs-Henseleit cycle is functional. In addition, a number of amines are excreted by the adult worm. These are presumably produced by the action of amino acid decarboxylases, which have been shown to be present in the tissues. There is some evidence that these various products of nitrogen metabolism are largely excreted by way of the gut. The gut of *Ascaris* appears to be an astonishing multifunctional organ.

Hemoglobins of *Ascaris*

Ascaris has two hemoglobins, one associated with muscle cells and one in the pseudocoelomic fluid. The muscle hemoglobin is half-saturated with oxygen at oxygen tensions of about 0.1 mm. This hemoglobin also shows a reversed Bohr effect.* Maximum dissociation of *Ascaris* muscle hemoglobin occurs at pH 7.0. It has been pointed out that this may be of adaptive value to an organism living under high carbon dioxide and low oxygen tensions, as in the vertebrate gut. The pseudocoelomic fluid hemoglobin, which is an octomer of hemoglobin units, has not been thought to function as a respiratory pigment. It is difficult to produce dissociation of this hemoglobin by lowering the oxygen tension, and dissociation is accomplished chemically only by strong reducing agents. This hemoglobin also has the highest oxygen affinity of any known hemoglobin. The suggestion has been made that it might function in facilitating oxygen diffusion. It has also been regarded as a metabolic pool for the synthesis of hemoproteins in the eggs.

Ascaris may be a microaerophilic organism. There is no question that oxygen is harmful when present at tensions resembling those of air. Laser showed some years ago that peroxides accumulate under these conditions and that *Ascaris* is apparently incapable of detoxyifying these peroxides. The characteristics of the hemoglobins described above appear to be adaptations for a microaerophilic life.

Neuromuscular System in *Ascaris*

Neuromuscular relationships in nematodes are peculiar. Rather than nerves sending processes to muscle cells, muscle cells send processes to the

---

* With vertebrate hemoglobins, an increase in hydrogen ion concentration shifts the oxygen dissociation curve to the left; that is, oxygen dissociates more readily under acid conditions.

longitudinal nerve cords or to the nerve ring. In *Ascaris* there are a number of individual muscle cells attached to each quadrant of the body wall. Each muscle cell consists of a fiber region that is contractile, a cell belly filled with glycogen, and an arm extending to a nerve cord and terminating in myoneural junctions. In the vicinity of the nerve cord each arm is subdivided into a number of terminal arborizations that intertwine with those of neighboring arms. This intertwining of terminal processes forms a longitudinal band parallel to, and in close contact with, the nerve cord. This band has been termed the *muscle syncytium*. These relationships are diagrammed in Fig. 6–16.

The adjacent fingerlike processes of the muscle syncytium have tight junctions between neighboring cells, whereas the neuromuscular synapses are characterized by a synaptic gap of about 500 Å. There are large mitochondria and clusters of synaptic vesicles within the presynaptic cytoplasm associated with the neuromuscular junctions. The fiber component of the muscle cell is made up of two types of myofilaments that interdigitate, forming H, A, and I bands as in vertebrate striated muscle. Rosenbluth considers the muscle cell of *Ascaris* to be intermediate in structure between the smooth and striated muscle types of vertebrate animals.

Because of the size and ready availability of *Ascaris,* almost all ex-

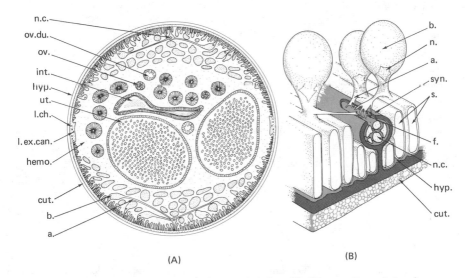

(A)   (B)

Fig. 6–16. Nerve-muscle relationships in *Ascaris.* (A) Cross section of female *Ascaris* to show the general arrangement of organ systems. (B) Details of nerve-muscle relationship. (B redrawn from del Castillo, 1969.) Abbreviations: a., arm, or innervation process of muscle cell; b., muscle cell belly; cut., cuticle; f., fingers or terminal branches of muscle arms; hemo., hemocoele; hyp., hypodermis; int., intestine; l.ex.can., lateral excretory canal; l.ch., lateral chord; n., nucleus; n.c., nerve cord; ov., ovary; ov.du., oviduct; s., muscle spindle (contractile); syn., muscle syncytium; ut., uterus containing eggs.

perimental work on the muscle and nerve cells of nematodes has been carried out with this worm. Acetylcholine is present in the worm, and it has a specific acetylcholinesterase. Several workers have found that acetylcholine produces contraction and that γ-aminobutyric acid (GABA) causes the relaxation of muscle strips. Both substances have been implicated as neuromuscular transmitter substances in invertebrates. Interestingly, the anthelmintic drug piperazine mimics the effect of GABA, causing the relaxation of muscle strips. When acetylcholine is applied by micropipette, no response is elicited in the belly or the fiber region of the muscle cell; however, depolarization occurs when acetylcholine is released on the muscle syncytium, showing that the receptors are restricted to this region of the muscle cell. The effects of acetylcholine are blocked by *d*-tubocurarine, although the rhythmic action potentials are not blocked. This is further evidence for the myogenic nature of these rhythmic depolarizations.

In 30 percent sea water and in some saline mixtures, more or less resembling *Ascaris* body fluid, the muscle cells have a transmembrane potential of -27.5 to -30 mV. Spontaneous polarizations occur at frequencies of 1.5 to 12 per second. The generation of action potentials in the somatic musculature differs from that in vertebrate striated muscle. In the latter, action potentials are generated by nerve impulse transmission across neuromuscular junctions. In *Ascaris,* action potentials are myogenic, arising from membrane activity of the muscle syncytium. In this way *Ascaris* muscle resembles vertebrate heart muscle. Nerve cord fibers appear to function in modulating myogenic action potentials through excitatory and inhibitory neuromuscular junctions. As already noted, acetylcholine and GABA may be involved as transmitter substances at these junctions.

Ascaris effectively regulates Cl^- in the perienteric fluid that bathes the muscle cells in the intact animal, and Cl^- appears to be important in maintaining the transmembrane potential of muscle. Data on the relative roles of K^+ and Na^+ in maintaining membrane potentials seem to be contradictory.

REFERENCES

Crofton, H. D. 1966. *Nematodes.* London: Hutchinson & Co. (Publishers) Limited.

Florkin, M., and B. T. Scheer (eds.). 1969. *Chemical Zoology,* Vol. 3. *Echinodermata, Nematoda, and Acanthocephala.* New York: Academic Press, Inc.

Goldschmidt, R. 1937. *Ascaris—the Biologist's Story of Life.* Englewood Cliffs, N.J.: Prentice-Hall.

Gould, S. E. (ed.). 1970. *Trichinosis in Man and Animals.* Springfield, Ill.: Charles C Thomas, Publisher.

Ratcliffe, L. H., H. M. Taylor, J. H. Whitlock, and W. R. Lynn. 1969. Systems analysis of a host-parasite interaction. *Parasitol.* 59:649.

Rogers, W. P. 1962. *The Nature of Parasitism.* New York: Academic Press, Inc.

Stirewalt, M. E. 1966. Skin penetration mechanisms of helminths. In *Biology of Parasites.* E. J. L. Soulsby (ed.). New York: Academic Press, Inc.

Thorne, G. 1961. *Principles of Nematology.* New York: McGraw-Hill Book Company.

Parasite Adaptations

chapter 7

Having discussed in modest detail several species of animal parasites, we may profitably consider in more general terms the adaptations associated with infectiousness, establishment in a host, and transmission to new hosts—three categories that we considered in Chapter 1 to be useful for the examination of parasitism. Since we have chosen these examples for examination in an arbitrary manner, mainly on the basis that we know more about them than about some other forms, it will be necessary to use some other examples from time to time in order to make clear certain points relating to parasite adaptations. Further, although the announced subject is parasite adaptations, it will be clear that the host must also be considered. In the preceding chapters the emphasis has been placed on the parasite in each case. However, as we indicated in the first chapter, the essence of parasitism lies in the fact that we are dealing with two reactive systems. Thus the reactions of the host are of great significance in considering the adaptations of the parasite.

ADAPTATIONS FOR INFECTIOUSNESS

We may define the infectious state as one in which the parasite can both *tolerate* the host and *react* to the environment furnished by a host in a fashion that will lead to establishment. In forms having more than one kind of host in the life pattern, there are separate infectious states, each of which is involved in culminating an association with a particular type of host. As an example, the specializations of the protoscolex of *Echinococcus* for tolerating the digestive juices of a carnivorous mammal are quite different from those of the shelled oncosphere of the same species, the latter tolerating the digestive juices of herbivores

144

but not those of carnivores. Similar adaptations are seen in tapeworms of the genus *Taenia,* which parasitize carnivores and herbivores at different stages of the life cycle. This means that there is marked physiological specificity in the particular infectious state. We saw a similar situation in the *brucei* subgroup of trypanosomes (p. 30). The stumpy forms of these trypanosomes are infectious for the arthropod vector, whereas the long spindly forms found in the salivary glands of a tsetse fly are infectious for the vertebrate host. Similarly, the miracidium of *Fasciola* is capable of infecting snails but not mammals, whereas the metacercaria is infectious for mammals but is unlikely to develop in molluscs.

In some trematode life patterns, a second intermediate host, such as a fish, frog, or arthropod, is involved in the life pattern. In such instances at least three different states of infectiousness exist in the life history of the parasite. The cercaria is infectious for the second intermediate host but is not capable of developing to sexual maturity in this host. An additional infectious state, the metacercaria, is attained in the second intermediate host. *Nanophyetus salmincola,* a trematode parasite of snails, fishes, and canines, is a good example of this more complex situation (Fig. 7–1). As might be anticipated, this is the pattern generally observed in the cases of trematodes utilizing a carnivorous animal as the host in which it attains sexual maturity. The adaptiveness of this arrangement should be obvious.

Parasitic animals in an infectious state reach the vertebrate definitive host by one of three general routes:

1. They are passively ingested along with food or water;

2. They are carried to the host by some intermediary organism, such as a biting fly or mosquito; or

3. They are attracted to and penetrate the host.

We may examine adaptations associated with these various routes.

Parasites that enter the host in food or drink immediately encounter an incredible, instantaneous modification of environment. They are thrown into a cavity, the stomach, which is specifically adapted for the denaturing of protein and its preliminary hydrolysis. It may be pointed out that the stomach, a peculiarly uncongenial environment for living things, undoubtedly serves a function in blocking the colonization of the metazoan body by an enormous assemblage of other organisms that are purposely or inadvertently ingested by animals. A parasitic organism entering such an environment must have adaptations that will allow it to tolerate the effects of hydrochloric acid or a change in osmotic pressure, and in mammals and birds it must tolerate an instant change in the temperature of the environment.

In many cases, cysts, shells, or sheaths that are resistant to the action

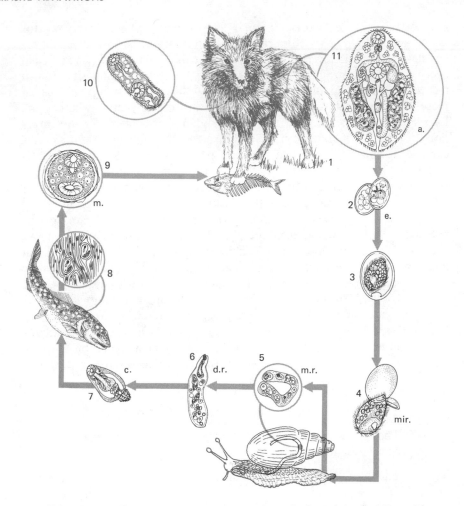

Fig. 7–1. The life pattern of the trematode *Nanophyetus salmincola*. The adult fluke (a.) lives in the small intestine of a fish-eating mammal, such as a dog or fox. Undeveloped shelled embryos (e.) leave the host in the feces, and development occurs in water. The fully developed miracidium (mir.) hatches and enters a snail. Here it metamorphoses into a mother redia (m.r.), which gives rise to daughter rediae (d.r.), which in turn give rise to a short-tailed cercaria (c.). The cercariae leave the host and on encountering a fish, typically a salmon, encyst as metacercariae (m.) in the musculature of the fish. Carnivorous mammals become infected by eating fish containing metacercariae.

of stomach juices enclose the infectious form.   Such structures include cysts of many protozoans, the cystic structures of infective tapeworm or trematode larvae, the shells of nematodes such as *Ascaris,* or the sheaths of infective nematodes such as *Haemonchus.*   In at least one case, a protozoan, *Histomonas meleagridis,* is protected from the stomach juices of the bird host by entering it while enclosed in the shell surrounding the infectious stage of the cecal nematode *Heterakis*.

A few forms seem to be able to tolerate the stomach without any obvious special sheaths or membranes. *Trichinella spiralis* and the migrating stages of *Ascaris* seem to have a remarkable tolerance for the stomach without any protective enclosing structures. Having no cyst to protect it, *Trichomonas hominis* seems to play a "numbers game," a fairly large inoculum being necessary to establish an infection in the intestine.

As will be clear, a parasite that has weathered the rigors of the stomach cannot remain huddled in its cyst or shell after it passes out of the stomach—if it is to infect the host. If it fails to react in some new way to the environment offered by the small or large intestine, the parasite will soon find itself again in the outside world. In the older literature the tacit assumption was often made that parasites were liberated in the host by the simple direct action of the host's digestive enzymes. However, study of several cases has shown that other components of the digestive tract juices serve as stimuli to activate a hatching or exsheathment response by the parasite. The role of carbon dioxide in the exsheathment of *Haemonchus* (p. 114), the excystment of *Fasciola* (p. 64), and the hatching of *Ascaris* (p. 131) have already been mentioned. Rogers has compared these effects to those of hormones. It should be mentioned that carbon dioxide does not act in a completely universal fashion as a stimulus for such responses. Another nematode, *Nematodirus battus,* utilizes pepsin-hydrochloric acid as a signal for the exsheathment response. In several protozoan and flatworm parasites, the salts of bile acids seem to activate infectious forms and induce excystment, although it has not been clearly shown that the bile salts stimulate secretion by the parasites of substances directly involved in their emergence from the enclosing structure. In many of the examples studied, elevated temperature, salt concentrations, pH, and oxidation-reduction potential modify the effects of the stimulus. Further, bile salts and trypsin appear to act synergistically in some cases.

There is some specificity in the action in vivo of the stimuli for exsheathment of infectious nematode larvae. Sommerville showed that the exsheathing stimulus was received by infectious larvae of various species at levels of the host gut anterior to the site at which each species characteristically develops into adulthood (Table 7–1).

Generally, parasites that are carried to a vertebrate host by a bloodsucking intermediate host are not capable of penetrating the unbroken skin. We shall deal with these presently in terms of adaptations for transmission.

The penetration of the outer covering of hosts by parasites involves processes that are essentially mechanical, but it usually also involves the action of chemical effectors. The parasitic plant nematode *Heterodera* has been shown to be capable of penetrating membranes by strictly me-

TABLE 7–1. Relation between the site of exsheathment of trichostrongyle nematode larvae and the site of their establishment in the intestinal tract of sheep.

Species	Exsheathment	Maturation
Haemonchus contortus	Rumen	Abomasum
Trichostrongylus axei	Rumen	Abomasum
Ostertagia circumcincta	Rumen	Abomasum
Trichostrongylus colubriformis	Abomasum	Small intestine
Nematodirus spp.	Abomasum	Small intestine
Oesophagostomum columbianum	Small intestine	Large intestine

After Sommerville, 1957.

chanical means. Dickinson described the process as occurring in three stages:

1. The nematode first adheres to the hydrophobic surface with the open mouth.

2. It then assumes a vertical position.

3. Penetration is finally effected by the extrusion against the membrane of a stylet in the mouth cavity (Fig. 7–2).

Combinations of mechanical and chemical processes used in penetration were described for *Schistosoma* (p. 74) and are known for several other forms. Alterations of basement membranes and of the ground substance have been demonstrated to occur during the penetration of skin by the nematodes, *Strongyloides* and *Nippostrongylus,* and in intestinal mucosa invaded by oncosphere larvae of the tapeworm *Taenia.* The miracidium of *Fasciola* produces cytolysis during penetration of the mollusc host. This is thought to be due to action of secretion from the so-called apical gland, which is emptied during the penetration process. Some nematode parasites of plants may utilize chemical aids in penetrating hosts. Nematodes of the genera *Ditylenchus* and *Pratylenchus,* for example, produce cellulolytic enzymes.

ESTABLISHMENT IN A HOST

Most animal parasites show rapid developmental changes after entering a vertebrate host. The specific parasites discussed in preceding pages undergo growth, and in most instances they also undergo differentia-

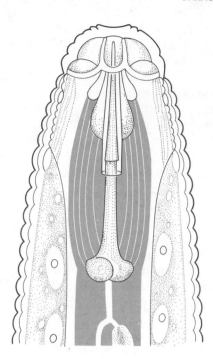

Fig. 7—2. The head of the nematode *Heterodera* showing the well-developed mouth stylet.

tion at the cell and organ levels. Clearly there are modifications of metabolism. Forms living in or out of the digestive tract often exhibit dramatic modifications of energy metabolism. It may be rationalized that the modifications of metabolism seen in *Ascaris* are those associated with the assumption of an anaerobic life. For example, those seen in the trypanosomes cannot be so interpreted, and the selection pressures that have led to such modifications of metabolism in the vertebrate host must be sought elsewhere.

Our information on the changes in the physiology of parasites during their early life in a host is somewhat fragmentary. The profound rapid modifications in sensitivity of schistosomes to saline, serum, or water during the rapid transition from cercaria to schistosomula (p. 75) suggest that other physiological changes occur, but these do not appear to have been investigated. Similarly, the excysted cysticercoids of *Hymenolepis diminuta* (p. 82) are rapidly killed by water, whereas the encysted forms survive for some hours under the same conditions. The cysticercoid metabolizes glucose at a very low rate, but immediately after excystment, the rate increases severalfold. Similar observations have been made on the acanthocephalan *Moniliformis*.

We know very little about the physiological changes in parasitic protozoans that take place during their early development in the vertebrate host, but it seems probable that changes do occur rather quickly. Newer culture methods for studying the events of differentiation in parasitic protozoans may yield information that has been difficult to obtain by harvesting parasites from hosts during the early phases of infection. As with other microbial agents, protozoans are difficult to recover for study in the period immediately after infection.

In recognizing that physiological modifications (undoubtedly related to later events of growth and differentiation) occur shortly after entering the vertebrate host, we should examine the question as to whether the direct stimuli for such modifications arise from the host or from the parasite itself. For example, those stimuli encountered on entry might set off a series of feedback reactions, and therefore no additional stimulus from a host would be required. There is some evidence on this point. It has been observed that the growth pattern of the tapeworm *Echinococcus* is determined by the physical character of the medium in which it is grown in vitro. In a liquid medium containing serum, the worm continues to grow asexually, producing new larval forms. This is its normal pattern of extraintestinal growth in a vertebrate intermediate host. On the other hand, if a semisolid protein substratum, such as heat-coagulated serum, is provided along with a liquid overlay, the worm undergoes strobilization and develops to sexual maturity. With an agar substratum and a serum-containing overlay, strobilization does not occur. Thus the growth pattern seems to be determined by the chemical *and* physical nature of the substratum with which the larval tapeworm is in contact.

The simplest physical stimulus for development is observed in certain tapeworms of the genus *Schistocephalus*. These forms undergo little net growth in the definitive bird host, but they do undergo rapid differentiation and subsequent maturation of the sexual organ systems. In vitro the major stimulus seems to be temperature. Semianaerobiasis and a container of a soft tubular character are also required for optimum development.

It has been suggested that mitochondrial growth in hemoflagellates is inhibited by temperatures at levels corresponding to those pertaining in a bird or mammal host, and that such growth is related to the changes in metabolism seen in the African trypanosomes and in some of their relatives. However, the finding that some individual hemoflagellates switch in the vertebrate to the type of metabolism seen in flagellates from the invertebrate host would argue against this (see p. 30).

The importance of carbon dioxide as a stimulus for the exsheathment of infectious larvae of *Haemonchus* has already been discussed (p. 114). However, there is evidence that carbon dioxide is also important as a stimulus for development from the third- to fourth-stage larva, a process normally occurring during the early period of establishment in the vertebrate host. If the sheaths of third-stage larvae are removed by treatment with weak hypochlorite solution and the larvae are incubated in saline, development to the fourth stage occurs in the presence of high concentrations of carbon dioxide. This development occurs in about 72 hours, which is about the same as the corresponding development period in the gut of a host.

Sites of Establishment

As we have already indicated, parasites living in the digestive tract show preferences for particular regions of the gut, depending on the species involved. Forms that live in the small intestine do not usually live in the large intestine. Even in the small intestine a particular region is typically favored for colonization. Interestingly, some parasites that show a rather high degree of localization when they are in the digestive tract may inhabit a variety of other organs if they escape from the intestinal environment. For example, the dysentery amoeba *Entamoeba histolytica* lives only in the cecum and colon when it inhabits the gut. If, however, this parasite invades extraintestinal locations, it thrives in the liver, brain, kidney, and even the skin. Similarly, the flagellate *Trichomonas gallinae* is restricted to the anterior part of the digestive tract as long as it remains in the digestive tract. When it spreads to sites outside the digestive tract, it may live in any soft tissue of its vertebrate host. We receive the distinct impression that the digestive tract imposes some serious constraints on the specific location in which a parasite may successfully establish itself.

The varying nature of the factors in the gut that may limit the region suitable for a particular parasite becomes apparent when we examine the known properties of the gut lumen. The stomach, of course, is an organ whose contents are at a low pH. Just below the pylorus the pH rapidly rises to values approaching neutrality, and there is a gradient of increasing pH from the pylorus to the posterior end of the small intestine. The chemical constitution of the small intestinal contents undergoes changes along its course. The most obvious changes involve the digestion and absorption of food, the secretion of large amounts of organic material by the glandular structures associated with the gut, the resorption of major portions of these secretions along the course of the gut, and a variety of chemical activities of microorganisms in the posterior part of the small intestine. Thus there are gradients in the concentrations of a variety of organic solutes of both dietary and body origin. Gradients in the oxidation-reduction potential of several vertebrate species have been demonstrated. Therefore it would appear that the gradients along the digestive tract, both chemical and physical, are greater in magnitude and probably involve a greater variety of variables than the differences encountered between any of the tissues outside the digestive tract. In addition to these linear gradients, there are gradients between the region in immediate proximity to the mucosa and the center of the lumen. The paramucosal lumen has a higher oxygen tension and may have higher concentrations of certain organic compounds than the central lumen.

In addition to these physical and chemical characteristics of the small intestine, the morphology of the mucosa seems to be important in some cases. Differences in the length of the intestinal villi or in the morphology of the rugae may determine whether or not a particular parasite can successfully attach itself in a region that is chemically compatible with its establishment. The size and configuration of the crypts of Lieberkuhn and of the lamina propria may also help decide whether or not an organism such as *Echinococcus* can become established. Thus we see that a variety of factors may be involved in determining the specificity of localization in the digestive tract of a host.

Many parasites that live outside the digestive tract also show some predilection for certain organs, but we have very little sound information on the bases for such localizations. In the past there has been a tendency to overemphasize the rigor of extraintestinal localization of parasites, in spite of the fact that it was known that organ specificity was not an absolute affair. For instance, the so-called lung fluke *Paragonimus* frequently turns up in organs other than the lungs, although it still occurs in the lungs with greatest frequency. Many other forms show this looseness in organ specificity. This may suggest that the behavioral responses of many parasites are not completely specific. We see the failure of responsiveness on the part of parasites in hosts that are not compatible with a completely normal development. The liver fluke *Fascioloides magna* that normally inhabits the bile ducts of deer behaves quite peculiarly when it is found in cattle. Instead of completing its migration to the bile ducts, it wanders in the parenchyma of the liver, creating bloody havoc in that organ. Similarly, the larvae of ascarid nematodes normally found in dogs and cats wander in the body if they get into a human host. They do not complete migration to the intestinal tract and never come to sexual maturity. This suggests that the human host does not offer the proper signals for completion of migration.

ADAPTATIONS FOR TRANSMISSION

Like many marine organisms, animal parasites must maintain high reproductive rates in order to beat the odds that the offspring will not survive to find a host and live another day. As we have seen from the examples chosen for discussion, the mechanisms used to enhance the reproductive rate are highly variable, but the end result is similar. The chances that a young parasite will reach a host in which it will in turn be able to reproduce are quite small. If it encounters a host, there is a certain chance that the host will be unsuitable for further development, or the host may already be suffering from disease or old age and may not survive long enough for the parasite to reproduce. Maintenance of a high reproductive rate seems to be a rule of the game, and parasites show great biological ingenuity in maintaining a favorable balance.

Among the protozoans very high reproductive rates seem to suffice by the relatively simple mechanism of a high fission rate. It might be argued that the combination of high concentrations of nutrient materials and elevated temperatures, for those forms living in warmblooded vertebrates, would explain the high reproductive rates seen in many trypanosome or trichomonad infections. Whether or not this explains the rates is unimportant in this context, since these organisms *do* reproduce at rates higher than those observed with their free-living relatives, and the biological function of improved chance for transmission is served.

The rather curious forms of multiple twinning or polyembryony seen in the life patterns of trematodes, such as *Fasciola* and *Schistosoma,* and in some cestodes, such as *Echinococcus,* have obvious advantages in insuring the perpetuation of the parasite species. A single larval trematode or tapeworm encountering a satisfactory intermediate host may give rise to hundreds or thousands of individuals. This seems to even things up for all those miracidia or oncospheres that do not find their way to intermediate hosts. Parasitic wasps also show the phenomenon of polyembryony. In these wasps the female lays a single egg in each host encountered. Since the female produces a number of eggs, the female can move quickly to another suitable host that is likely to be in the vicinity and lay another egg. Each egg undergoes polyembryonic development, giving rise to a number of new wasp individuals in the intermediate host. This enhancement of reproductive potential is important because most females of these wasp species fail to find suitable hosts in which to deposit their eggs, and such females die or abort the eggs.

In the case of tapeworms, high reproductive rate in most species is maintained by the replication of the reproductive organs along the strobila of the animal. Some authors have referred to the strobilate tapeworm as a colony and, in a sense, it is. Self-fertilization seems to occur with high frequency in cestodes, although there is some evidence that continuing self-fertilization in *Hymenolepis nana* is deleterious. At any rate the production of eggs in the strobilate forms of tapeworms results in the production of a very large number of potential tapeworms. *Hymenolepis diminuta* may produce up to 250,000 eggs per day throughout the life of its rodent host. This is a potential for the production of about 100×10^6 young tapeworms in the host's life. If all of these reached hosts and attained maturity as strobilae, it would represent over 20 tons of tapeworm tissue. Clearly this does not occur, and the chances must be very great that a *Hymenolepis* oncosphere will not be eaten by an intermediate host.

Hermaphroditism not only occurs in the cestodes that are often self-fertilizing, but it is also a common pattern among the digenetic trematodes. As a matter of fact, *Schistosoma,* which was chosen as an exemplary trematode in preceding pages, is unusual in having separate sexes. It is not known how often self-fertilization occurs among the digenetic trematodes, but reproductive potential is enhanced by having every individual in the definitive host capable of producing both eggs and sperm. Parthenogenesis occurs among some parasitic nematodes but is not particularly associated with parasitism, since it also occurs among the free-living nematodes.

In many cases the rate at which eggs are produced by parasites is extremely high in forms that do not show polyembryony, hermaphroditism, or strobilization. A female *Ascaris,* for example, will produce 200,000 eggs per day. As was once said, these parasites don't eat to live; they eat to reproduce. Thus we may conclude that high rates of reproduction are almost the rule among parasitic animals, and plainly these high rates are adaptations for transmission.

The adoption of various forms of periodicity in parasites is another form of adaptation for transmission. Some of the filarial nematodes, such as *Wuchereria,* show a peculiar form of diurnal periodicity. These worms are transmitted from vertebrate to vertebrate through the biting of mosquitoes. The young worms appear in the peripheral blood of the vertebrate host on a periodic basis in some localities. The time of appearance always corresponds to the preferred biting hours of the vector. In some of the Polynesian islands the vectors show no particular time of the day as a preferred biting time, and in those localities the worms show no periodicity. The human pinworm, *Enterobius,* exhibits a regular periodicity, migrating to the rectum and anus of the host during the hours in which the host sleeps. Pinworms living in individuals who sleep in the daytime undergo migration in the daytime, so it seems clear that the migration is associated with a physiological state of the host.

There are instances in which periodic behavior on a seasonal basis is manifested. In *Leucocytozoon,* a relative of the malaria parasites living in birds, there is a very well-developed seasonal periodicity. This is closely correlated with the activities of the intermediate host, which is a biting fly. This periodicity may be triggered by the endocrine state of the host. The protozoan parasite of frogs, *Opalina,* shows a seasonal periodicity in the production of cystic forms that leave the host and may infect new hosts. The initiation of sexual reproduction and of cyst formation in the protozoan is correlated with the level of gonadal hormones in the host and results in the liberation of infective forms of the parasite at a time when new young hosts are available for infection. One of the most astonishing periodic patterns in a parasite is that of the rabbit flea *Spilopsyllus.* Ovarian maturation in the flea is stimulated by an increase in the blood corticoids of the pregnant host. Shortly after young rabbits are born, the fleas move from the doe to the young rabbits in the nest. The level of the pituitary growth hormone in the young rabbits stimulates the fleas to copulate, and shortly thereafter the fleas lay eggs. About ten days after the birth of the young rabbits most of the fleas return to the lactating doe, and complete regression of the fleas' gonads occurs.

Cleveland and his associates have studied one of the most interesting endocrine relationships between a host and protozoan inhabitants of the gut. There are a variety of flagellates living in the wood roach *Cryptocercus punctulatus*. When the roach is undergoing periodic molts, a clear synchronization of the sexual cycles of the protozoans takes place with the molting of the host. It was shown that the sexual cycles of the flagellate protozoans are triggered by the production of the molting hormone, ecdysone, in the host. As a matter of fact, the protozoans are more sensitive to the hormone than is the host itself. Hormone levels that do not result in molting induce sexual cycles in the protozoans. Of the several genera of protozoans involved, some are more sensitive than others.

Parasites that are carried to a new host by an intermediate host, such as a fly, a flea, or a biting bug, typically enter the new host through the skin, although such parasites cannot penetrate unbroken skin. In many cases the parasite exhibits adaptations that result in its being injected into the host while the intermediate is obtaining a blood meal. In such cases the pattern of development in the invertebrate host involves migration of morphologically distinguishable forms, such as sporozoites of *Plasmodium* (p. 11) or metacyclic trypanosomes (p. 26), to the salivary glands or to the mouthparts of the invertebrate. Essentially nothing is known of the stimuli that may trigger this pattern of morphogenesis and migration. In at least one group—the filarial nematodes, the worms complete their development in the cephalic portion of the hemocoele in the insect host (a fly or mosquito), and they actually break out of the mouthparts when the intermediate feeds on a vertebrate. They may enter the host through the bite wound produced by the biting arthropod.

Some forms do not migrate to the mouthparts or associated structures. The trypanosome *Schizotrypanum cruzi* completes its development to the infectious state in the posterior gut of a biting bug, *Triatoma*. It depends for transmission on the contamination of the bite wound or other break in the skin by the feces of an infected bug. Not surprisingly, the efficiency of various species of *Triatoma* in transmitting *Schizotrypanum* is related to the frequency with which their feces are deposited while they are feeding on vertebrate hosts.

In each of the cases mentioned above, the pattern of development in the intermediate host is closely linked with the mode of transmission.

Those parasites that reach and penetrate a host without "aid" from another host might be expected to have host-finding adaptations. A number of such adaptations are known. For example, the golden nematode

Heterodera rostochiensis, a parasite of potato plants, reacts to substances diffusing into the soil from host roots. The activating material stimulates the emergence of larval *Heterodera* from cysts and their migration to the roots. The stimulating substance is a cardiac glycoside. The larvae of several skin-penetrating nematodes are known to show positive responses to temperature, and schistosome trematodes (p. 74) may react to the warm surface of a host.

A number of trematode cercariae show responses to light interruption or water turbulence, and such responses would be of significance in bringing them into contact with fish hosts. Trematode miracidia show responses to chemical factors emitted by snails (p. 73); however, these appear to be quite nonspecific. Miracidia of *Fasciola* or *Schistosoma* react to a great variety of snails, many of which will not support subsequent development of the parasites.

A related type of response is seen in some of the filarial nematodes living in the subcutaneous tissues of vertebrates. When the arthropod vector feeds on the vertebrate, it injects salivary secretions into the bite wound. The young infectious stages of the worms, known as *microfilariae,* are attracted by the secretions and rapidly move to the site of the bite wound. This markedly enhances the chances that the worm will be taken up with the blood meal of the feeding insect.

It should be pointed out that the reactions elicited in parasites by chemical substances associated with hosts are not a class of reactions peculiar to organisms that live as parasites. A number of such reactions to what are termed *pheromones* are known to serve in prey-finding, mate-finding, and other such social relationships among free-living animals. However, such adaptations in parasites serve roles in transmission.

Even in closely related species of parasites, the stimuli that lead to development of an infectious state suitable for transmission may vary sharply. The trypanosomes furnish good examples of such differences. *Trypanosoma mega,* a parasite of African toads, was found to convert from the epimastigote to the trypomastigote stage when treated with serum of toads or mammals. Subsequently it was found that the same effects could be produced by treating the cells with urea. Not all cells in a culture were capable of making the transformation, and maximum competence coincided with the onset of the stationary growth phase of cultures. Thus the ability to change from epimastigote to trypomastigote seems to depend on two factors, one of which is intrinsic and perhaps time-dependent, the other being extrinsic and supplied by a host. The addition of urea to a *T. mega* culture inhibits the uptake of tritiated thymidine by transforming cells, but does not affect those cells in the population that lack competence for transformation. It is not possible to say whether this difference in the uptake of the nucleic acid precursor represents a cause or an effect of the transformation.

Temperature may play an important role in the developmental cycles of some trypanosomes living in warm-blooded hosts. The invertebrate stages of trypanosomes from mammals can be cultured on a variety of media at 26 to 28° C. When such cultures of *Trypanosoma lewisi,* a parasite of rodents, are exposed to temperatures of 30 to 31° C, most of the cells are killed. By using very complex media, several workers have been able to maintain trypanosome stages of *T. lewisi* and its close relatives *T. conorhini* and *T. theileri*. When the insect host was used as a source of inoculum for *T. theileri* cultures at 37° C, trypanosome stages such as occur in the vertebrate host developed in the cultures. Thus, with this group, temperature seems to play an important role, although its action is not completely understood.

In addition to temperature it is clear that the chemical constituents of the medium are important in allowing the development of the trypanosome stages. This might be related to the observed changes in nutritional requirements that appear with increased temperature in the case of *Crithidia fasciculata*. This organism is a close relative of the trypanosomes and lives in the gut of mosquitoes. When its nutritional requirements are studied at different temperatures, they were found to undergo dramatic change with increases as small as 0.2° C. Compounds that stimulate growth but are not essential at one temperature become absolute requirements for growth at slightly higher temperatures (Table 7–2). Thus the

TABLE 7–2. Changes in the factors required by *Crithidia fasciculata* with increases in the temperature of cultivation.

Temperature	Requirement
22–27°C	Minimal defined medium supports good growth. Stimulation by increased amounts of leucine, isoleucine, phenylalanine, and tyrosine.
28–30°	Factors stimulatory at 22–27° C are now essential. Stimulation by glutamate, succinate, lactate, and additional Ca^{++}, Fe^{++}, and Cu^{++}.
32°	Factors stimulatory at lower temperatures are now essential.
32.4°	Additional carbohydrate is required.
33–33.5°	1.5 × concentration of whole medium, sufficient at last temperature interval, is stimulatory.
33.6–34°	Additions of lecithin, choline, and possibly of vitamin B_{12} and inositol are stimulatory. An extract of dried egg yolk is more stimulatory than the defined compounds.

After Guttman, 1963, Exptl. Parasitol. 14:129.

interaction between temperature and nutritional requirements may be quite important in the change of a hemoflagellate from life in an invertebrate host to life in a warmblooded vertebrate host.

SPECIFICITY OF PARASITE HOST

In nature a given parasitic species is most often found in a quite limited number of host species. In the most extreme cases, distribution of a parasitic species may be limited to one kind of host. In many instances the host distribution is clearly related to the interlocking of the life pattern of the parasitic species with those of potential hosts. For example, the distribution of a parasite that develops in fish-eating animals may be dependent on the fish-eating proclivities of birds and mammals for completion of its life cycle, and thus fish-eating will be a broad determinant of the host distribution of the parasitic species. The trematode *Nanophyetus* (p. 146) is a case in point. Similarly, a form with a life pattern such as that of *Fasciola* (p. 59) is most likely to be distributed in grass-eating animals, as indeed it is. A modification of the feeding pattern of a host may lead to the acquisition of new parasites or to the loss of parasitic species by that host. We see this most readily in the case of man. The distribution of the beef tapeworm *Taenia saginata* in man clearly depends on his eating undercooked fresh beef. In countries in which uncooked or undercooked fish is eaten, man serves as a host to parasitic species that use fish as intermediate hosts.

Parasites seem to be capable of developing in many more species of hosts when under laboratory conditions than when under actual conditions in nature. Sometimes this may be due simply to opportunity as indicated above. In other cases this is not a complete explanation. The dwarf tapeworm *Hymenolepis nana* is commonly found in certain murine rodents in nature. In the laboratory it will infect a large number of host species, including the gray squirrel *Sciurus carolinensis,* if the hosts are fed on ordinary laboratory animal diets. If, however, *Sciurus* is fed on acorns and mushrooms (its "natural" food), it will not support the growth of *H. nana,* and it is highly refractory to infection. Similarly, a trematode found in mollusc-eating birds will develop in ducks only if these hosts are fed a diet containing molluscan tissue. Obviously in such cases the parasitic species is adapted to hosts that have restricted dietary habits and are affected by the chemical composition of the diet itself.

There remains of course the possibility of physiological incompatibility between a parasite and its potential host. This might involve the failure of the host to supply the stimuli appropriate to establishment in a host. However, as we have seen (p. 147), these appear to be stimuli of a very general quality. On the other hand, if stimuli do not act rapidly enough, the parasite may find itself on the way out of the host before the events of establishment are initiated. For example, the mouse can be infected with *Hymenolepis diminuta* quite readily if the host has received a drug that depresses intestinal motility, thus increasing the time required for intestinal emptying. Otherwise, some strains of mice appear to be relatively refractory to infection by this tapeworm. This relative refractoriness seems to be due to *H. diminuta* being carried through the intestinal tract too rapidly to allow easy establishment. Similarly, the trichostrongyle nematodes might fail to develop if the exsheathment stimuli were received at a place in the digestive tract posterior to sites allowing satisfaction of the worms' requirements for growth and reproduction (see Table 7–1, p. 148). Thus the timing of stimuli may be at least as important as the quality of stimuli leading to establishment.

Physical or chemical incompatibility may also involve conditions that result in death of the parasite. For example, several parasites normally developing in insect hosts fail to develop in the American cockroach *Periplaneta* because they are apparently destroyed by the grinding action of the proventriculus, or some tapeworms normally developing in herbivores are killed in the stomach of carnivores. Whether the latter is due simply to retention in a highly acidic environment is not known, although this has been offered as an explanation. The high concentration of urea in the spiral intestine of elasmobranch fishes seems to serve as a barrier to colonization of this habitat by some worms. Several helminths from birds and mammals have been shown to be killed by similar concentrations of urea (250–500 mM), and some tapeworms found in elasmobranchs have been found to require urea in the external medium for maintenance of osmotic balance. Similarly, protoscolices of *Echinococcus,* a tapeworm developing to sexual maturity in carnivorous hosts, are killed by the bile salt glycocholate but not by taurocholate. Glycocholate is a predominant bile salt in herbivorous animals, as taurocholate is in carnivores. On the other hand, the shelled oncosphere of *Echinococcus,* which undergoes subsequent larval development in herbivorous animals, is killed by intestinal fluids or bile of dogs and cats. White mustard and some other cruciferous plants are not susceptible to infection by the nematode *Heterodera rostochiensis* because the roots contain allyl thiocyanate, which kills nematode larvae.

Clearly the specificity of parasites for hosts is not attributable to just one class of parameters, but rather it may be a complex phenomenon requiring analysis in any given case.

HOST RESISTANCE TO PARASITES

Although a given host may be immune to a parasite by virtue of being constitutionally unsuitable as a habitat (p. 158), it is important to recognize that the establishment of a parasite in a host also involves the capacity of the parasite to resist *reactions* of the host that might be deleterious. These host reactions would include inflammatory tissue responses and immune responses of a humoral character. Both types of host reaction occur in most parasitisms involving a vertebrate. However, such reactions are usually not rapidly sterilizing in their effects on parasites. In trichinosis, for example, it has been shown that the life span of the adult worms in the host is markedly affected by inflammatory reactions in the intestine of the host. Drugs that suppress the inflammatory response, such as cortisone, prolong the life of the adult worms. However, the worms normally reach sexual maturity and deposit a large number of young worms in the host's tissues before the host responses terminate the worms' reproductive careers. In some instances, such as schistosomiasis, the immunity is incomplete, and worms continue to grow to sexual maturity and may live for a period of years, although the number of new worms that are successful in establishing themselves in such an "immune" host may be reduced. In most animal parasitisms, immunity is not of the solid sterilizing type seen in some viral or bacterial infections. This incompleteness of immunity is related to the long-term debilitating effects of many animal parasitisms involving human and domestic animal populations.

For a long time many parasitologists who were interested in immune reactions were prone to regard immunity to animal parasites as different on a physiological basis from immunity to bacteria or viruses. However, the preponderance of evidence now indicates that the basis of immunity is the same with all infectious agents in a given host, although the functional immunity of the host that may be displayed varies from case to case. The differences seem to be related to the fact that immunity is not a single reaction of a host but a complex multivariable affair. The interplay of these variables will determine the nature of the total immune response of a host to a given "foreign" agent. A brief review of mechanisms may clarify the difficulties in understanding a given immunity.

The immune system in higher vertebrates is composed of several tissue components. These include fixed cells of the spleen, liver, lymph nodes, and thymus. There is also a system of mobile phagocytic cells, originating mainly in the bone marrow and including the heterophil and eosinophil leucocytes. In addition, there are a number of potential macrophage cells in the body, such as undifferentiated mesenchyme cells. Some of these cells have the ability to elaborate immune globulin proteins or antibodies, which may be released from cells into the body fluids. These globulins may combine with parasitic organisms or their products and produce an inhibitory effect, or they may render the parasites susceptible to attack by phagocytic cells.

The first reaction of a vertebrate to a foreign agent may be a direct reaction of fixed macrophages. If the entering parasite is small enough to move about in the blood, the parasite may rapidly be phagocytized. In the case of some parasitic organisms the response of host cells works to the parasite's advantage, and the phagocytosis may not function as an immune response at all. The phagocytosis of malaria sporozoites, for example, simply results in the parasite reaching an intracellular position in the host, where it then proceeds to multiply. In the case of larger parasites entering the tissues there may be destruction of host cells. This releases histamine, which causes an attraction of wandering macrophages and sets up what is a primary stage of inflammation, and, if it occurs rapidly enough, a parasite may be markedly inhibited. Such a reaction occurs most often when a parasite enters a host in which it does not usually develop. In some parasitisms this reaction is quite mild, unless the host has been previously infected with the parasite. We shall return to this last point shortly.

The second phase of an immune response involves the reception of information by lymphoid cells. The presence of a foreign substance (antigen) is "recorded," and in response, cells of the lymphoid-macrophage system elaborate antibodies.

The immunoglobulins appear in various molecular forms, perhaps arising in different ways from different cells. They not only vary in molecular weights and other physical properties, but they also vary in action; some pass rather readily across membranes, and some attach readily to the outside of host cells. The macroglobulin (IgM) is the first to appear in an infected animal. There is some evidence that this antibody absorbs easily to phagocytic cells and promotes phagocytosis of antigens (or of parasites). This is followed by the appearance of gamma globulin (IgG) and alpha globulin (IgA). The proportion and number of antibodies belonging to these categories vary with individual parasitic infections. Although it is tempting to speculate that these forms of globulins serve specifically different functions in immunity, the evidence for this is quite incomplete. In many infections involving animal parasites there is frequently a proliferation of eosinophil leucocytes, resembling the response seen in allergy. In a few cases it has been shown that the host is indeed in a hypersensitive state and that this may prevent superinfection by a new inoculum of the parasite (p. 116). In some other instances the resistance to superinfection does not clearly involve allergy. In many protozoan infections, phagocytosis by fixed macrophages is characteristic of superinfection resistance. In some worm infections there is enhanced infiltration of the tissues around the parasite by eosinophils, Mast cells, and lymphocytes from the blood.

In malaria infections there is increased phagocytic activity, particularly in the spleen, against parasitized red cells as well as against free parasites. This results in destruction of many of the parasites and is a reasonably effective immune response, as long as the host is infected. If the host is cured or loses the infection, this immunity quickly disappears. However, while infected, such a host is very effectively protected against new infection by the same strain of the parasite. This type of immunity, sometimes called "premunition," is very specific, since no protection against a new strain of the parasite is effected by the presence of a different strain in the host. Further, the rapid loss of immunity may allow the rapid development of susceptibility in a population, resulting in cyclic outbreaks of acute malarial disease. There is little evidence that a free circulating antibody is also involved in malaria immunity.

In schistosomiasis, part of the immune response seems to be against the shelled embryos that are deposited in the tissues of the host. These responses are allergic in nature and are responsible for a number of the symptoms seen in schistosomiasis. However, if an animal has previously been infected with schistosomes, it shows some resistance to infection when there is a new inoculum of the parasite. A number of antibodies against schistosome antigens have been demonstrated in the blood of infected animals, but there is also evidence that some of these antibodies play no part in a functional immunity of the host. This functional immunity against schistosomes is incomplete in that it is not sterilizing in effect. An infected host tends to be resistant to the establishment of new worms after the initial infection, but the worms already in residence are little affected (see p. 165).

Mention should be made of the possible role of substances called "interferons" in the immunity of hosts to animal parasites. These substances are produced in response to the presence of a foreign agent. Unlike antibodies, they seem to render host cells unfit as habitations for intracellular parasites. It has been shown that interferons may operate in malaria and in some other protozoan infections.

Late manifestations of immune response are seen in a number of cases. For example, the calcification of encysted *Trichinella* that occurs after some months in the host or the fibrous capsule that forms around the hydatid cyst appear to be resistance responses. Such effects are always manifested by a preliminary infiltration with wandering cells of the lymphoid-macrophage system.

There are other types of host response that cannot be clearly recognized as immune in character. For example, the changes in the fatty acids found in the digestive tract of animals infected with the tapeworm *Hymenolepis,* or the marked modification of the amino acid composition of bile in animals harboring *Fasciola hepatica,* are obviously host responses to the presence of the parasites concerned. It would be of great interest to determine whether such responses are beneficial to the parasites, to the host, or to neither.

BEHAVIOR AND RESISTANCE

In laboratory experiments with mice and voles, Christian found that with increases in population there were increases in the weight of adrenal glands and decreases in the weight of thymus glands. The changes in any individual animal were shown to be related to its position in the social hierarchy or "peck order" of the population; those highest in the social hierarchy showed the least change. Increased adrenal weight is accompanied by an increase in the secretion of adrenal corticoids, and this in turn suppresses inflammatory responses and phagocytosis. Davis and Read showed that the intensity of muscle infection by *Trichinella* in house mice was related to their position in the social hierarchy and to their adrenal weight, "low-caste" animals having the highest worm burdens. It has thus been postulated that social behavior and number in the population would aid in the control of population size by regulating the immune responses to infectious agents. Parenthetically, in the case of trichinosis, this phenomenon would also regulate the number of infectious agents acquired by a predator feeding on a rodent population. This might act as a feedback damper on a predator population.

This example has been mentioned to demonstrate the ecological complexities involved in the control of host-parasite populations in contrast to the simple interactions that may pertain in an isolated host-parasite system.

REFERENCES

Christian, J. J., and D. E. Davis. 1964. Endocrines, behavior, and population. *Science* 146:1550.

Dogiel, V. A. 1966. *General Parasitology.* New York: Academic Press, Inc.

Faust, E. C., P. C. Beaver, and R. C. Jung. 1968. *Animal Agents and Vectors of Human Disease,* 3rd ed. Philadelphia: Lea & Febiger.

Fisher, F. M., Jr. 1963. Production of host endocrine substances by parasites. *Ann. N.Y. Acad. Sci.* 113(1):63.

Olson, W. O. 1962. *Animal Parasites: Their Biology and Life Cycles.* Minneapolis: Burgess Publishing Co.

Read, C. P. 1970. *Parasitism and Symbiology.* New York: The Ronald Press Company.

Rogers, W. P. 1962. *The Nature of Parasitism.* New York: Academic Press, Inc.

Evolutionary Considerations

chapter 8

It is evident that the forms we have chosen to discuss as exemplary animal parasites have adopted this way of living along quite independent lines of evolution. Although the two trematodes discussed may have had a closer common ancestry than some of the other forms, it is also apparent that there has been a considerable evolutionary divergence among the trematodes after the assumption of a parasitic life pattern. This leads us to the often stated idea that the assumption of parasitism is an evolutionary blind alley. Although it may be true that the assumption of this mode of life imposes constraints on the path of evolution that may be followed, it is also quite clear that parasitism opens up a large array of new evolutionary opportunities for organisms. The tapeworm genus *Hymenolepis* furnishes a fine example of the exploitation of opportunities. Speciation has occurred repeatedly in this group, and several hundred species of *Hymenolepis* are now known from among avian and mammalian hosts. There is reason to believe that the tapeworms are an ancient group of parasites in vertebrates, and that the evolution of the group has occurred in tandem with the evolution of hosts. In a sense it may be said that the evolution of the parasites in this group has been dependent on the evolutionary opportunities of their hosts.

It is widely accepted that particular kinds of hosts are associated with certain kinds of parasites, and that such mutual relationships have arisen and been stabilized over long periods of time. We have also pointed out that there is a certain amount of host organ specificity that can be observed in many animal parasites. In some instances this may have led to a rapid isolation of mutants from the parent stock without the mutant entering a new host. Again, the genus *Hymenolepis* furnishes examples that may represent this type of isolation. There are a number of instances of separate hymenolepid species living in the same host individual among bird and insectivore hosts. We know little of the specificity manifested by these forms in the host intestine. Schad has shown that the pinworms of tortoises may inhabit quite different niches in the intestine of the host, so that a form of microgeographical isolation occurs.

Many aspects of parasite evolution may be treated in terms not significantly different from those utilized in the study of free-living organisms. However, there appear to be some special selection factors (or lack thereof) acting on parasite populations, and some special mention should be made of these.

IMMUNITY AND EVOLUTION

In the preceding pages something was said of the immune responses of the host. In most of the parasitisms discussed, some immune responses are known. Generally they are not sterilizing in effect; that is, the responses do not result in the eradication of the parasite population in the host. However, recent studies have suggested that immune responses may be of significance in parasite evolution.

In the case of metazoan parasites, Dineen has pointed out that genetic continuity demands that the parasitic organism remain in contact with the host until the attainment of reproductive capacity, and that this period of required contact may exceed the period required for an effective immune response by the host. Thus the selection pressure of immune responses will favor the perpetuation of those variants that have reduced antigenic dissimilarity from the host. In the selection of variants exhibiting lessened antigenic differences from the host, parasites may show (1) deletion of genes for antigenic properties, (2) acquisition by mutation of antigenic properties of the host, or (3) both of these.

Damian showed that hosts and schistosome trematodes share antigens and referred to this sharing of constitutive antigens by host and parasite as "molecular mimicry." Presumably, molecular mimicry could evolve by change of host or parasite genotype. However, the briefer life and shorter generation time of the parasite make it more likely that major selection would occur in populations of parasites. French workers, notably Capron and his colleagues, have regarded the common antigenic properties of schistosomes and hamsters to be genetic. However, relying heavily on inducible enzyme theory, they argued that the specifically similar antigens are induced in the parasite as it grows in a particular host. Apparently the major difference between Damian and the French workers seems to depend upon the question of whether the antigenic similarities are constitutive or inducible. From the standpoint of the evolution of genotypes this difference is not significant.

Recent research by Smithers and his coworkers seems to be consistent with the hypothesis of Capron et al. When schistosomes that had been grown in mice were transplanted to monkeys, the worms ceased egg production and became shrunken in appearance. After five to six weeks, however, the worms recovered, and egg production was resumed at a normal rate. On the other hand, when worms were transplanted from mice to monkeys that had been immunized previously against mouse spleen or red cells, the worms died in less than 44 hours. When worms were transplanted from mice to monkeys and after several days transplanted to monkeys previously immunized against mouse tissues, the worms survived. Immunological studies showed that the worms from mice contained mouse antigens, and these observations also indicated that the worms grown in mice assumed some antigenic properties of mice, properties that might be lost after a few days in a host of different species.

However, it boggles the imagination to assume that schistosomes contain antigenic information for a large number of specific antigens found separately in mice, monkeys, gerbils, hamsters, as well as man. Applying Occam's razor, it seems more likely, as suggested by Smithers and his colleagues, that the worms may incorporate host proteins, particularly in or on the tegument, and thus escape recognition by the host's immune mechanisms. Such an ability would require genetic information, but it would be considerably less complicated than coding for a great array of proteins from different hosts. Further, it is very unlikely that *Schistosoma mansoni* has had significant evolutionary experience with the laboratory mouse. Clearly more research is required for a full understanding of the antigenic similarities assumed by parasites in their hosts.

In a number of parasitisms, hosts respond to the parasite by producing a number of antibodies that do not adversely affect the survival of those parasites in the host. This may be explained by the assumption that the parasite that has had "experience" in the host has developed the antigenic similarity discussed above, and although such a parasite may induce immune responses in the host, it is shielded from their effects by its cloak of similarity. On the other hand, new parasites entering such a host have not had experience with the host and are not protected from the immune mechanisms of the host. This may explain the protection against new infection from schistosomes seen in hosts that already harbor an infection. This would be an important aspect of population regulation.

When two or more species of parasites are present, antibodies against one parasite may affect another parasite. This would be of significance in affecting the competition of two parasite species infecting the same host species. For example, previous infection of mice with *Schistosoma mansoni* results in some immunity to infection with another blood fluke

Schistosomatium douthitti, and *Schistosoma bovis* infections seem to confer some protection against *S. mansoni* infections. This is a form of competition that is peculiar to parasites.

Thus the evolution of parasites may be affected by immune selection
factors, although investigation of such parameters has been undertaken
only recently.

WHAT DO PARASITES HAVE IN COMMON?

Having examined a heterogeneous assemblage of parasitic organisms in preceding pages, we must next consider whether any common patterns emerge. There is little evidence of convergence in the morphology
of these forms. The nematodes appear to remain round and the flatworms
to remain flat. However, a question arises as to whether there are any
physiological characteristics held in common. The obvious one that leaps
to the mind is the tendency for an abbreviation of metabolism. The loss
or modification of the now traditional tricarboxylic acid cycle is an obvious
case in point. Accompanying this, there is also a general pattern of ex-
creting partially oxidized products of energy metabolism. Carbon is incompletely oxidized. In addition to this particular characteristic there
may be a general requirement for carbohydrate as an energy source.
Further, a remark should be made concerning the general significance of
an absolute nutritional requirement for carbohydrate. Free-living animals, with few exceptions, can utilize fats and proteins as energy sources.
In most instances a properly balanced intake of fat will satisfy the energy
requirements of most free-living animals. However, the parasitic animals
generally are incapable of utilizing fats as energy sources when living in
the vertebrate host. This seems to be commonly true and is not surprising
in view of the general pattern of excretion of fatty acids as major end
products of energy metabolism.

The patterns of energy metabolism that can be termed abbreviated
are obviously accompanied by the loss of biosynthetic capacities associated with the pathways that are missing or repressed. Hence another
general attribute of these organisms appears. Energy metabolism in free-
living animals is closely linked with biosynthesis at the level of carbon
skeleton manipulation; that is to say, energy metabolism furnishes certain
carbon skeletons for the synthesis of carbon compounds. In examining
the energy metabolism of parasitic animals, it is striking to note that the
input of carbon can often be accounted for in terms of carbon excreted by
the organisms. In short, energy metabolism appears to be connected
with biosynthetic function at the level of energy transfer, probably involving the synthesis of adenosine triphosphate, but seems to have very little
connection in terms of the supply of carbon skeletons. Catabolism and

anabolism appear to be almost separated in these organisms. This may be a striking difference between the free-living and the obligately parasitic animals. More data on this point are needed.

What other types of specialization can be associated with this assemblage of organisms? In looking at all of them, we recognize that in the host the parasite has available to it quantities of "food," which are limited only by the capacity of the host to supply them. In common with the cells of the host itself, the parasite has available to it all of the carbon compounds and all of the physical properties of the host fluids. These chemical and physical characteristics are of course under the control of the host, and we would say, if we were speaking of the host itself, that they are regulated by homeostatic mechanisms of the host body. Thus the parasite has available to it all of the effects of the control mechanisms that serve the host in maintaining its integrity as an organism.

In some cases there is good evidence that the parasitic phases of animal parasites have deleted or suppressed control mechanisms that are essential to many free-living organisms. For example, the parasitic phases of the animal parasites discussed in preceding chapters cannot osmoregulate. Further, it has been shown that a number of animal parasites exhibit no negative feedback in a thermal gradient. They migrate in such a gradient until they reach a temperature level that is lethal. Free-living animals move in a thermal gradient until a given temperature level produces sufficient negative feedback to slow and eventually halt net progression; in other words, we would say that the organism has attained its temperature optimum. In parasitic animals the host furnishes a thermoregulatory device or moves to a place where the temperature is conducive to continued life.

In the case of the forms that live an intracellular life, such as the malaria organisms, regulation includes those cellular mechanisms concerned with the maintenance of the host cell itself. We have already mentioned the fact that the malaria organism may lack the ion gradient-coupled systems that characterize cells in general. All the evidence available would suggest that the animal parasites are primarily parasitic on the homeostatic mechanisms of the host, and this is a characteristic they have in common. This is a considerably more significant common property than a requirement for some exotic compound the supplying of which is only a manifestation of the dependence on the homeostatic mechanisms. Thus measurements of the flow of carbon compounds from host to parasite and from parasite to host should be examined in terms of regulatory mechanisms.

In addition, some mention should be made of the capacity of the parasite to elicit a response from the host. In the case of *Trichinella,* a special point was made of the fact that the parasite seems to produce a change in the muscle fiber, which results in this structure being modified into a "house" for the parasite, rather than continuing to function as a muscle fiber. Similarly, in the case of *Hymenolepis,* the worm stimulates the host to change its pattern of fatty-acid secretion into the lumen of the gut. In this latter case, disease is not manifest; all efforts to demonstrate pathology in *H. diminuta* infections of the rat have failed. Thus we have an interlocking of metabolisms: the parasite acts, the host reacts, the parasite reacts, the host reacts. A new arrangement of the homeostatic mechanisms occurs, and new properties characteristic of the parasitism appear. In the case of a number of these parasitisms it has been shown that there are modifications of the energy metabolism of the host. This is often not discernible by crudely measuring total metabolism, but it can be detected when, for example, the deposition of glycogen in the liver is examined. Evidences of the modification of host metabolism have been found in all cases in which the biochemistry of the host has been examined in sufficient detail.

In view of the fact that the parasitic animals seem to excrete partially oxidized products of energy metabolism, the statement has sometimes been made that metabolism is "wasteful." This is of course an anthropocentric view of the organisms, based on the notion that getting all the energy out of a compound has some intrinsic merit. It is plain that for these organisms such a theory does not have special merit. Rather, we might expect that the production of partially oxidized compounds would have some biological advantage. If the ancestors of these organisms carried out more complete oxidations of energy sources, the subsequent evolution of the groups should have favored those individuals having the capacity to carry out complete oxidations, assuming that the organisms lacking these capacities had no selective advantage or disappeared through natural selection. Clearly this latter assumption does not seem to be the case. It has been argued that the patterns are those we would expect of organisms living an initially anaerobic life in the gut of the host. However, there is no reason to believe that this is the explanation for the abbreviations of metabolism seen in the trypanosomes. In some of these there seems to be an advantage for the parasites in repressing the capacity for complete oxidations while they are living in the blood of the vertebrate host. The selection pressure for incomplete oxidation is not obviously associated with anaerobiasis in the trypanosomes, since the pathways for terminal oxidation include the reduction of oxygen.

Possibly there is a selection for abbreviated energy metabolism in parasitic organisms. If it is remembered that the selection pressures experienced by free-living organisms include those associated with the food-getting activities and all of the complex activities that may be related to food-getting, as well as the fact that food may be available quite intermittently and irregularly, it may be rationalized that there is positive selection pressure for the maximum oxidation per unit of chemical energy source.

On the other hand, these selection pressures do not operate strongly on the parasitic organism. Rather, the selection pressure would mainly be related to the reduction of work functions not directly associated with the reproduction of the organism. This would include the maintenance of systems for RNA and for protein synthesis not required in reproductive function. In the presence of a constant food supply, an organism in which the energy produced in metabolism is not required for food-getting and corollary functions would be favored by natural selection. This would lead to the abbreviated metabolism seen in the parasitic organisms. In these terms the application of the words "wasteful" or "inefficient" is inappropriate.

Similarly, the loss of biosynthetic capacity might have selective advantage for organisms living in a host. Lwoff suggested some years ago that the saving in energy effected by the deletion of capacity for biosynthesis might favor those individuals showing such deletions. Zamenhof and Eichorn examined this experimentally in a bacterium and found that auxotrophs indeed had a selective advantage over prototrophs, provided that the required substance was present in the medium.

In the case of the trypanosomes or of *Ascaris* there is clearly repression of certain metabolic systems during life in the definitive host. Such repressions would be energetically less advantageous than deletions, since DNA and RNA synthesis are still cost factors for the organism. However, it may be shown that the limitations of food supply in other parts of the life cycle can act as a powerful selection pressure to maintain these systems. In *Ascaris* the larval stages are sealed in a shell, and metabolism is restricted to those substances that are enclosed in the shell by maternal metabolism. Under these circumstances there would appear to be a selection pressure favoring complete oxidation of the limited food available. In the case of some trypanosomes the life of the organisms in the gut of the intermediate host may mean a shortage or irregular supply of an energy source. Thus repression in the vertebrate host is evidently favored, but deletion is not. It should be pointed out that those trypanosomes in which genetic deletion of the information for the Krebs cycle may have occurred have also dispensed with life in an arthropod gut as a part of the life cycle and have evolved other mechanisms for transmission to new vertebrate hosts, such as venereal or mechanical modes. However such genetic deletion of the Krebs cycle has not been demonstrated.

SIGNALS FOR DEVELOPMENT

The development or life history of an organism is an orderly affair. The information in the zygote nucleus is expressed as a series of events with feedback from the cytoplasm. Further, as development proceeds, the semiautonomous elements of the cell, such as mitochondria, act upon and are acted upon in what is clearly a systematic fashion. As the multicellular organism grows, interactions between cells and between groups of cells produce further orderly modifications of intracellular interactions, and we recognize the events of differentiation at the multicellular level. Clearly the development of an organism involves the reading out of information sets with complex feedback mechanisms. We may glibly make such a general statement, although our understanding of the detailed mechanism is quite imperfect.

If we examine the life patterns of the specific parasites discussed in preceding pages, we are struck with the fact that at one or more points in development, chemical and/or physical information must be furnished *by a host* in order for development to proceed. If such information is not furnished by a host, development of the parasite does not proceed. If we think of development as a series of closed feedback loops, we should recognize that among free-living organisms there are instances in which there appears to be an open loop that can be closed by an environmental factor such as temperature. In the case of parasitic organisms the occurrence of open feedback loops in development is the rule. These loops are closed by chemical and physical cues derived from a host. These cues for parasite development have a uniqueness in that they are derived from information bits in the genotype of the host organism.

In discussing the role of carbon dioxide in stimulating the exsheathment of worms such as *Haemonchus,* Rogers has compared the phenomenon to an endocrine system. The host furnishes specific chemical information to a receptor. This, perhaps through neurosecretory mechanisms, triggers the secretion of exsheathment fluid, causing liberation of the parasite in the host. As we have seen, chemical signals trigger the excystment of *Moniliformis, Fasciola, Hymenolepis,* and protozoans parasitizing the vertebrate intestine, as well as the hatching of various shelled nematode larvae. The specific chemical stimulus may vary from one group of species to another. It is important to understand that we are speaking primarily of signals or stimuli that may be of very short duration but nevertheless are critical in triggering the initiation of further development.

Intensity of the signal may also be important. Can we, in fact, distinguish the changing nutritional requirements during development from what we have termed the closing of an open loop? It is perhaps important to make such a distinction. Heterotrophic organisms, whether free-living or parasitic, require a certain number of preformed carbon compounds that are derived from other living things. Thus any developing organism that requires a carbon compound from exogenous sources may show an arrest of growth, and ultimately death, if the substance is not available.

Smyth has speculated on the significance of changing levels of nutrition during the phases of the parasite life cycle. This might be expanded to include other environmental features. For example, in addition to a regular external supply of carbohydrate as an energy source to drive growth, tapeworms such as *Hymenolepis* require a temperature of about 38° C for growth of the strobilate phase and the maintenance of a temperature below 32° C for normal development of the cysticercoid. Similarly, the carbon dioxide requirement for *development* of *Haemonchus* is probably explicable as a chemical role of carbon dioxide in energy metabolism. This environmental requirement can be distinguished from the signal stimulus of carbonic acid that triggers exsheathment of the third stage larva of *Haemonchus*. In the latter case, *Haemonchus* larvae may be exposed to carbonic acid for a short period of time and the stimulus removed; exsheathment ensues in the absence of carbonic acid. These environmental requirements do not appear to operate as signals to restart the sequences of development but rather are required components of development.

There are other instances in which it is not yet clear whether a signal or a required component for development is involved. It will also be obvious that whereas a signal may be required to close an open feedback loop, the ensuing development must occur in an environment that is chemically and physically suited to sustain developmental events. Thus a set of rather tight constraints may be involved in the completion of development.

The general necessity for a signal from the host that "tells" the parasite that it is in a host and triggers a set of physiological responses is associated with the adoption of parasitism, and by combining a second, or even third, simple chemical effector of host origin, the response may be made quite specific and sensitive. The present author has referred to this elsewhere as a principle of interrupted coding; that is, there is a block in the genotype information of the parasite. The host furnishes a product representing this block and thus allows further development at the point in time that is suitable for further development. This appears to be a general attribute of animal parasites and may pertain in other parasites as well.

A GENERAL REMARK

In the preceding pages an attempt has been made to describe the biology of some selected parasitic animals. Rather than a case history approach, we might have treated animal parasitism in the context of symbiosis, as originally defined by de Bary in 1879. De Bary's definition concerns itself with the degree of integration existing between the members of a symbiosis rather than the reciprocity of benefits presumed to be derived by the reacting species. In some measure the degree of integration and the resulting characteristics of the symbiote host would depend on the overall environmental context. Symbiosis would represent a unique biological entity, but it would not necessarily involve a rigid biological "obligation" of the organisms in association. Using this approach, we might escape the use of terms such as *mutualism, commensalism,* etc., which are often used as though they represented legal contracts or treaties between friends or enemies. Several years ago Gregory suggested the use of the term *functional field* as a substitute for the terms used to classify different categories of symbiosis. This would of course allow the framing of questions concerning the degree of symbiotic integration as a function of the intensity of the field established and of the resistances to its establishment.

Biologists are accustomed by habit to talking of natural processes as random in character and in this way emphasizing the idea that nature operates in a statistical fashion without favoring one entity or state of affairs over another. To the earlier evolutionists, the concept of randomness in nature became almost a fetish in their zeal to avoid the traps of predestination or teleology. However, it is important to introduce and to maintain the limiting concept of adaptiveness, as it relates to the randomness of nature. When two entities such as a host and a parasite become adapted to one another and the combination begins to work as a system, selection against the random elements occurs. Nonrandom boundaries for the association appear, and the system develops tight constraints. The random elements in the association may for many practical purposes disappear, or at least be suppressed. Whitlock and his colleagues have pointed this up in studying haemonchosis and have shown that by directing attention to the boundaries, the system may become amenable to analysis. This is a very important consideration when we are observing the biology of a parasitism as it exists in nature, in contrast to its maintenance in the laboratory.

The study of parasitism in terms of boundary conditions requires the application of proper statistical methods for handling data and, more importantly, for determining what data are worth acquiring from the system. The sheer quantity of trivial data, available for the taking but shedding no light on the operation of the system, can fruitlessly consume the life of an investigator. The point to be made here is that the proper mathematical tools are essential to the investigator who would seek to supply a systems approach to problems of parasitism. The difficulty is that the individual must also become familiar with the biology of the system at the level of natural history.

REFERENCES

Dineen, J. K. 1963. Antigenic relationship between host and parasite. *Nature* 197:471.

Fairbairn, D. 1970. Biochemical adaptation and loss of genetic capacity in helminth parasites. *Biol. Rev.* 45:29.

Schad, G. A. 1966. Immunity, competition, and natural regulation of helminth populations. *Amer. Naturalist* 100:359.

Smithers, S. R., and R. J. Terry. 1969. The immunology of schistosomiasis. *Adv. in Parasitol.* 7:41.

Zamenhof, S., and H. H. Eichorn. 1967. Study of microbial evolution through loss of biosynthetic functions: Establishment of defective mutants. *Nature* 216:456.

Although we have not discussed the treatment of parasitic diseases in this book, the selection of examples for discussion was to a considerable extent dictated by a desire to choose forms about which we really know something. Because of the historical medical and veterinary interest, the best known forms are typically those implicated in the diseases of man or his domestic animals. For the most part the forms we have discussed produce diseases for which there is a well-developed chemotherapy, as well as other regulatory measures. Some of the most serious of these diseases, such as malaria, can be controlled. Some, such as the African trypanosomiases, can be treated but are not yet readily controlled. With the proper investments of time and money, it now seems probable that all of the diseases in which animal parasites are involved can be brought under some reasonable measure of regulation. In some instances the education of human populations would result in functional health improvement. In other instances, such as malaria, it has been shown that education is not essential.

However, there are some side effects from the control of infectious disease that have become the basis for serious and almost overwhelming problems, particularly in the disequilibration of world populations. With the control of malaria in certain areas of the world, we have witnessed quite astonishing effects on the dynamics of human population growth, and we have not yet been able to develop rational working mechanisms for population control.

Previously, in primitive societies, in addition to the heavy toll of life exacted by infectious disease, particularly malaria, which prevented the attainment of reproductive age by a significant proportion of the population, man had a variety of behavioral controls on population size. These

Epilogue

included ritual sacrifice, headhunting, cannibalism, castration, infanticide, and rigorous rituals for entering adulthood and marriage. These practices are of course unacceptable to modern man, and the new taboos have since been transmitted to almost every corner of the earth.

Historically, then, in many underdeveloped countries that had a high incidence of malaria, the populations had been maintained in the face of a high death rate by the maintenance of a high birthrate. When malaria control was implemented and death rates fell, the high birth rates continued. Consequently, some of the populations concerned have undergone explosive growth. In Mexico, for example, it is anticipated that the population will double between 1960 and 1980. Even more alarming is the prospect in a country such as Pakistan. Between 1920 and 1965 the death rate in Pakistan declined from 40 to 29 per thousand, and at the same time the birthrate of 55 per thousand has been maintained. To these statistics should be added the fact that Pakistan has one of the lowest living standards in the world, with an annual per capita income of only $80. The prospects for the future there are indeed frightening, and there is no easy way to transmit the necessary wisdom to the Pakistani people, especially since 80 percent of them cannot read. The formation of the new nation of Bengla Desh does not change the balance of people and available food. This balance may even be worsened. Another example is Ceylon where the control of malaria is credited with reducing the death rate by three-fourths in a single decade. This was accompanied by a 35 percent increase in population in the same ten-year period.

The above are illustrations of the effects of malaria control in certain selected areas. They are not quoted to show that the control of malaria is just around the corner, but rather to show that the control of the disease in man immediately results in the appearance of wholly new problems involving human welfare.

Thirty years ago malaria was common in all the temperate subtropical and tropical regions of the world, with the exception of a few islands. In the 1940s there were about 350,000,000 cases of malaria in the world, and about 3,000,000 people died each year from the disease. By the beginning of the 1960s the number of cases in the world was only about 100,000,000 with less than 1,000,000 deaths. This indeed represents a reduction, but not of the magnitude that is technically possible. Malaria still disables more people on earth than any other disease. With the increased awareness that malaria control is feasible, few governments can resist the pressure to bring it under control. In India, for example, it is estimated that the country loses about $400,000,000 per year from the effects of malaria. It is further estimated that it would cost about $114,000,000 to eradicate the disease in India. However, the real problem is what the cost to India would be of the increase in population that would result as an immediate side effect of malaria control. This is more difficult to evaluate, but it is worth consideration, especially in view of the fact that the already bulging population of India cannot be supported by the agricultural production of that subcontinent.

This is not to suggest that the control of malaria should be withheld from a people just because it might result in new and inconvenient problems. It does, however, make the point that we must take all factors into consideration when we apply manipulations of nature that may grossly upset the equilibrium state of a population. The population we are really talking about is the population of the world. In this age we can no longer use the fang-and-claw approach to problems involving human beings. A worldwide approach is needed. What happens in the Indian subcontinent will have an effect even on the American working man who sits in front of his television set and watches the news while he has a beer. Furthermore, the language we use in discussing disease problems suggests an attitude toward nature that may not be appropriate for the future. For example, we speak of the "conquest" of disease or of the "war" against disease. This terminology is a reflection of the notion that a man in his superiority can conquer nature, and it is a notion that we may have to abandon if we are to survive as a species.

Thus the management of malaria on a world scale must take into account all of the other side effects that might result if we treat it as a problem in a vacuum. Like deer, mice, or chestnut trees, human populations may have been somewhat self-regulating in the past, but by the use of nonheritable adaptations, such as chemotherapy or insecticides, mankind quickly outruns the natural processes operating in populations of plants and animals. Therefore it has become obvious that we must develop a new restraint in tinkering with nature if we are to avoid destruction of our habitable environment and to prevent our own extinction.

Index